D0256795

UNITED WE ARE UNSTOPPABLE

UNITED WE ARE UNSTOPPABLE

*60 Inspiring Young People
Saving Our World*

EDITED BY AKSHAT RATHI

JOHN MURRAY

First published in Great Britain in 2020 by John Murray (Publishers)
An Hachette UK company

I

Copyright © Akshat Rathi 2020

The right of Akshat Rathi to be identified as the Author of the Work has been
asserted by him in accordance with the Copyright, Designs and Patents Act 1988.

Speech on page 99 © Ashley Torres 2020
Speech on page 131 © Holly Gillibrand 2020
Letter on page 150 © Laura Lock 2020
Diary entries on page 213 © Zoe Buckley Lennox 2020
Speech on page 245 © Carlon Zackhras 2020

Maps and internal illustrations © Naomi Wilkinson 2020

A CIP catalogue record for this title is available from the British Library

Hardback ISBN 978-1-529-33594-1
eBook ISBN 978-1-529-33596-5

Typeset in Bembo by Hewer Text UK Ltd, Edinburgh
Printed and bound in Great Britain by Clays Ltd, Elcograf S.p.A.

John Murray policy is to use papers that are natural, renewable and
recyclable products and made from wood grown in sustainable forests.
The logging and manufacturing processes are expected to conform
to the environmental regulations of the country of origin.

John Murray (Publishers)
Carmelite House
50 Victoria Embankment
London EC4Y 0DZ

www.johnmurraypress.co.uk

CONTENTS

CONTENTS

INTRODUCTION

On Monday, 20 August 2018, fifteen-year-old Greta Thunberg sat down outside the Swedish parliament with only an armful of flyers and a wooden sign that read *Skolstrejk för Klimatet* (*School Strike for Climate*). She posted some photos on social media, but few paid attention.

The next day she sat down in the same place and went on strike again. A few people joined her this time. She continued to strike for twenty-one days until the Swedish general election, with more and more people protesting alongside her. Greta's story spread on social media, inspiring other young people across the world to take a stand against climate change.

On Friday, 15 March 2019, a global school strike was called. Seven months after Greta's lone protest, more than 1.4 million people took part, in 2,000 cities and towns in 128 countries, from Argentina and Australia to the UK and the US. The message to world leaders was loud and clear: act now on climate change before it's too late.

And they're right to demand urgent action. In October 2018, the United Nations' Intergovernmental Panel on Climate Change (IPCC) published a landmark report which concluded that, without radical action by 2030, there is little chance of avoiding the catastrophic impacts of climate change. To get there, the world will need to take unprecedented action to cut carbon emissions quickly.

We've known about global warming since the 1860s, when it was discovered that greenhouse gases such as carbon dioxide can warm the planet. Science tells us now that a hotter planet makes life worse for most living organisms: from unbearable heatwaves to flash floods and more intense hurricanes to mega droughts.

In the 1910s, reports warned that the burning of fossil fuels such as oil, coal and gas increases the concentration of these gases in our atmosphere, thus identifying human activity as a contributor to global warming. But the convenience of fossil fuels proved too tempting to abandon. Easy access to unlimited energy brought riches to Western countries, and provided plenty of excuses to push away worries about a distant future.

By 1987, there was little doubt that humanity's addiction to fossil fuels would lead to horrific and far-reaching consequences. Scientists predicted that rising sea levels would eat up entire islands and vast coastal regions. Extreme weather events would force hundreds of millions of people to migrate, creating a refugee crisis of epic proportions. And, in an evil twist of fate, those who would suffer the worst impacts would be people in poor countries that have contributed least to the greenhouse gas problem.

In the past three decades, the urgency of addressing the problem has grown but the actions taken have been nowhere close to proportionate. The first COP – the annual United Nations climate conference – happened in 1995, creating a framework for countries around the world to work together to create programmes that would cut emissions. But it wasn't until 2015, at the twenty-first meeting of COP in Paris, that all the countries in the world finally agreed on a goal to keep the rise of global average temperatures 'well below' 2 °C and pursue efforts to limit it to 1.5 °C.

Although the Paris climate agreement has begun to bring disparate groups together under one umbrella of climate action, the world continues to hit new annual records on carbon emissions. And with every passing year the world eats up its finite carbon budget – the amount of CO_2 emissions that can still be emitted before we blow past the targets set in 2015.

Today's young generation will inherit this planet and they know they are going to suffer the worst consequences of climate change, if real action is not taken now.

Greta Thunberg has become the face of the youth climate protests. She leads with steely determination and speaks with supreme clarity. She deserves all the praise that has come her way and more that will surely come. But she's also the first person to shun it and let others know that she isn't the leader of the youth climate movement.

Speaking at the 2019 UN climate summit in Madrid, she called for the media to turn their attention to the other young people fighting the climate crisis across the world: 'We are privileged and our stories have been told many times, over and over again. And it is not our stories that need to be told and listened to.'

Greta's story is one of many. This book is about those many. In the pages that follow, you'll read about sixty young people, in forty-one countries, who are fighting to save our planet on every continent.

There's Cricket Guest from Canada, protesting against the building of oil pipelines through indigenous land; Aditya Mukarji in India, who is changing how businesses operate one plastic straw at a time; Vivianne Roc in Haiti, working to ensure women's voices can be heard in the climate fight; Shannon Lisa, a chemical detective in the

US who recognises that extreme events linked to climate change could cause highly toxic chemicals to leak into the environment; Carlon Zackhras in the Marshall Islands, trying to raise the alarm about losing his home to rising sea levels; Lesein Mutunkei, a high-school student in Kenya who plants a tree whenever he scores a goal playing football and is convincing hundreds to join in, and many more who are proving that no one is too small to make a difference.

These young people don't just bring new energy to the climate fight; they bring new perspectives, fresh tactics and unwavering resolution. They don't just understand that everything in the world is connected; they also know how to bridge the divides that have been forming. They know that tackling climate change requires cutting emissions, but that getting there will require facing up to and rooting out deeper injustices perpetuated in societies.

The youth climate movement has sprung up from the grass roots, brought millions into the fold and changed the global conversation. Even as the world faces a different global threat in the Covid-19 pandemic, seeing these young people continue protests online, rebuffing attacks and accusations while remaining firmly grounded in science and humility, is proof that the movement is likely to endure.

Climate change will affect everyone around the world for decades to come, and that's why it needs a sustained, global movement to fight it. Greta's strike began as a one-person protest – it was people telling her story that encouraged many others to join the fight. The stories that follow will surely inspire many more.

Akshat Rathi
April 2020

ASIA

CONTINENT: ASIA

POPULATION: 4.4 BILLION

BIGGEST CLIMATE CHALLENGES

- **LARGE COASTAL POPULATION** – Sea levels could rise between one and three metres by 2100. To put this in perspective, two-thirds of Bangladesh is less than five metres above sea level. By 2050, one in seven Bangladeshis and more than 640 million people in South East Asia could be displaced.

- **MORE INTENSE STORMS** – Asia's large coastal population is vulnerable to extreme weather events such as typhoons and cyclones, which are expected to intensify as a result of climate change.

- **WATER SCARCITY** – The west of Asia is one of the most water-stressed regions on the planet. Between 1998 and 2012, the eastern Mediterranean Levant faced the worst drought of the past 900 years. In a hotter world, the likelihood of droughts increases.

- **GLACIER MELT** – Rising temperatures are melting glaciers, particularly in the Himalayas, which supplies water to the world's biggest rivers such as the Yangtze, the Ganges and the Indus. More than a billion people rely on the Himalayas' stores of fresh water for survival.

ADITYA MUKARJI

Aged 16
INDIA

In 2017, I saw a video of a sea turtle having a plastic straw painfully extracted from its nostril. It haunted me and I began to read anything I could on how this could be prevented. I wanted to make a difference. I wanted to help bring about measurable change and also spread awareness about the problem.

Recycling plastics should be the last option as it gives people a false sense of doing good while continuing to use more and more plastics. In any case, plastics cannot be recycled infinite times. Ultimately, they will end up in landfills or oceans, polluting our planet.

My campaign #RefuseIfYouCannotReuse is against the use of single-use plastics in everyday items like plastic straws, cutlery, crockery, water bottles and containers, as well as for packaging material like shrink-wrap for bananas and other fresh produce. I targeted the hospitality sector as it uses the largest number of single-use plastics items.

When I started, I focused on plastic straws because they are possibly the most redundant item ever created. Initially, businesses were reluctant to give up using plastic straws, because they feared customers would be unhappy. So I asked them to put up signs that showed the business's commitment to the planet. Plastic straws would be offered if customers asked. Next I suggested they use eco-friendly straws, and I sourced suppliers who could provide them. So far, my efforts have replaced over 26 million plastic straws and millions of other single-use plastic items.

I was particularly surprised by the amount of research material and knowledge I got from my father, who has worked for global companies like DuPont and Shell. It makes me believe that the industry knows the problems they have caused, but they do not have the will to change. It is up to us as consumers to use the power of refusal to educate ourselves and others.

* * *

As a teenager my main job is being a student just in case the planet survives and I need a career. But I am living with the climate crisis. Bad farming practices have caused the water table to drop to dangerous levels. The burning of rice stubble in northern India has polluted the air and is causing irreparable damage to citizens, including my family.

As one of the world's most populous countries, India will be one of the most adversely affected by climate change. The country seems ill-equipped to handle the effects. We are still struggling with issues of poverty alleviation. That's why limiting people's aspirations through government regulation is a political minefield. I would like to improve the quality of life of those at the bottom of the economic pyramid and close the gap between the desperately poor and the disgustingly rich. To do that, India will have to look at development along with the environment and not at the cost of the environment. It calls for a level of statesmanship that seems to be missing.

I cannot do anything about these issues, apart from joining climate protests to draw our government's attention to this problem. Unless we all join hands and push governments to regulate and consumers to change their use patterns, we are going to see a bleak future.

HTET MYET MIN TUN

Aged 18

MYANMAR

In May 2008, Myanmar was hit by Cyclone Nargis, which was the worst disaster the country has ever faced. Thousands of people lost their lives and millions became homeless. At the time, I was about six. I remember witnessing the cyclone with my family from the small window of my house: the wind uprooting trees and tearing rooftops apart. Although there were only a few casualties in the city of Yangon, where I live, compared to other delta regions, in the aftermath of the cyclone we witnessed scenes no one had ever imagined.

Even today I can hear the sound of the wind in my ears when I think about that tragedy. When I was older, we learned about climate change in the general science class; about its causes and its effects, deforestation, greenhouse gases and global warming. Connecting this knowledge to what I had seen of Cyclone Nargis, I started to realise that we have to take action to combat climate change before it is too late.

I am campaigning for more public awareness of climate change. I believe that if you know the type of disease, it is much easier to find the cure. In Myanmar, where people still have to fight for fundamental rights such as political rights, education and health care, there isn't much awareness of climate change and its consequences. I believe the public should be aware of two important things: climate change is now violating human rights; and everyone has a responsibility to protect the environment as a global citizen.

There are many challenges to being a youth activist, but the biggest is that we do not have many resources: money, information or networks. We are in need of more platforms, both locally and internationally, where youth activists can widen their networks and share problems, solutions, information and experiences.

* * *

In the future, I want to become a policymaker. For a developing country like Myanmar, the risks of climate change are high. There have already been predictions that climate change is likely to cause the flooding of coastal cities, the mass migration of millions of people across borders and scarcity of food and water resources. If these predictions come true, it will not be an easy task for our government, which lacks a strong economy, financial resources and advanced technology, to handle it. There are also risks that these vulnerable situations could easily escalate into civil unrest, which would in turn disrupt regional stability and security.

The sector most affected by climate change in Myanmar is the agricultural sector, on which the country's economy is heavily reliant. The monsoon season is becoming less predictable and the number of extreme weather events is increasing – cyclones, unusually heavy rainfalls and droughts are damaging both the quality and quantity of crops being produced.

These unprecedented extreme weather events are also affecting the well-being of many people in Myanmar; for instance, droughts cause shortages of water supply in many regions in summer, and floods and landslides in the rainy season displace thousands of people every year.

Although Myanmar is not a big contributor to causing climate change, it is one of the countries that are most vulnerable to its effects. Coastal cities, including Yangon, the commercial centre of Myanmar, are under threat because of sea level rise. The country's huge delta region will suffer too, because seawater will contaminate ground-water, destroy farmlands and erode land in many towns and villages.

There are three things I want to urge the leaders of my country to do. First, give climate change the importance it deserves. Collaborate with experts, scientists and activists to combat it. There are some things we the public can do individually, but for the rest we need government support, no matter how hard we try. Second, balance economic development with environmental protection. While Myanmar is opening its doors to the world and accelerating industrialisation, it is important not to do it at the cost of harming the environment. Our future generations deserve to drink fresh water, breathe fresh air and enjoy the beauty of nature. Third, formulate concrete strategies and emergency measures in order to prepare ourselves to face the potential threats of climate change. We lost thousands of lives when Cyclone Nargis hit. We cannot afford such a disaster again.

I urge every person, every organisation and every nation not to take advantage, including political advantage, of climate change. Climate change should not divide people, parties and nations. Instead, it should unite people of different professions and expertise, no matter where you are from, what you believe or what political party you support. I believe, with the collaborative effort of different nations, people and organisations, we can overcome the challenge of climate change and achieve the greatest success humankind has ever accomplished.

To succeed, we need four things: knowledge, wisdom, perseverance and determination. In solving climate change, we have the knowledge and wisdom; we already have the technical solutions. What the climate challenge needs now is perseverance and determination.

TATYANA SIN

Aged 26
UZBEKISTAN

From a young age I realised what a huge impact human activity can have on nature. I come from a region close to the Aral Sea. Well, it's close to what is left of it. I face the consequences of this tragedy every day: contaminated water, soil and air, and the deterioration of people's health.

My mom is a director of a local non-governmental organisation called KRASS that is devoted to helping local people and farmers to get by and live in our contaminated environment. She is my inspiration. The mission of her organisation is to contribute to the improvement of rural livelihoods and poverty alleviation as well as increasing long-term food security and environmental sustainability in rural Uzbekistan. If I hadn't witnessed her hard work and especially the positive outcomes that her actions bring, I would not have chosen this path. It is hard to stay indifferent when it comes to helping people.

Following her example, from my early college years, I've tried to work for environmental organisations. As an intern at UNESCO's Tashkent office, I developed a clear idea of how the consequences of natural disasters and climate change can be tackled. But I decided that rather than coping with the devastating consequences of human actions, I'd like to prevent them. Later, during my MBA, I worked on developing corporate social responsibility strategies for Uzbek companies. My aim is to improve the sustainability of the agricultural sector.

The main problem related to climate change that Uzbekistan is facing is the melting of glaciers, which is the only source of fresh water for the country. The economy of Uzbekistan is centred on the agricultural sector. This means that we are highly dependent on the water supply, the main source of which are the glaciers of the Tian Shan mountain range. If we don't cut emissions to stem global warming, the livelihoods of those working in agriculture will be threatened and the local community could lose its only source of fresh water. The constantly increasing temperature and water shortages will drastically worsen already poor living conditions.

* * *

The young generation is the future of this planet. That is why it is important to raise climate literacy among young people and motivate them to take action. Educational institutions must make young people aware of environmental problems and teach them not only how to adapt to climate change, but also how to avert it. Let's follow Italy's example and make learning about climate change and environmental problems compulsory at every level of education.

Climate change is the unfortunate reality of the modern world, but it is young people who can change it. More and more young leaders are taking action to raise awareness and prevent the human activities that cause climate change. It is the responsibility of governments to give young people a chance to speak up at local, national and global levels about environmental issues.

The Aral Sea was once the fourth largest inland lake in the world. But high water demands, particularly from the cotton industry, has led to a water reduction of 90 per cent. This has exposed large parts of the seabed, and led to huge salt storms that have been linked to an increase in respiratory diseases in those living nearby.

WHAT WOULD YOU ARE WORRIED ABOUT

Being concerned or worried about the climate crisis is significant in itself. Change begins because of people who care. Make changes, no matter how small. Do what you can, and even better, push yourself to do things that you think you can't do.

Akari Tomita, 16, Japan/USA

I know how scary it is. I know how it feels like there's nothing more you can do. But we still have time to make amends. I urge you to use your voice and contact your local politicians, make the message of the severity of the climate crisis known in your locality, and strike with us.

Theresa Rose Sebastian, 16, India/Ireland

SAY TO PEOPLE WHO
THE CLIMATE CRISIS?

It is natural to be worried. Climate change is an existential threat. Become a member of a group or organisation that advocates for action on climate change because having conversations on the ways to reduce the impacts of climate change can help reduce some of the stress.

Nasreen Sayed, 27, Afghanistan/USA

IMAN DORRI

Aged 28

IRAN

Protesting is one way to show dissatisfaction with the current situation, but I firmly believe that it is not the best measure to combat climate change. Even if we tell people that the world is in peril, some of them just think about their own short-term interests and do not want to accept the situation. So if we are going to allocate time to protesting against climate change, we should then allocate twice that amount of time to finding solutions that are both practical and can satisfy the majority.

That is why the primary focus of my activity is universities, urging them to be one of the pioneers in climate action and encouraging collaboration between students and faculty members.

Iran is located in an arid and semi-arid region of the world. I have seen first-hand how climate change is affecting people – from droughts to flash floods. These issues were the main reason why I began a master's in civil environmental engineering, where I learned about sustainable development. After graduation, I went on to work for the university's office of sustainability.

We need more people who talk less and act more.

Although the consequences of flash floods have been both severe and destructive in recent years, drought is the main problem Iran has been confronting as a result of climate change, and it is predicted to get worse. I live in Tehran, the capital of Iran, where, because of better infrastructure, the effects of climate change aren't felt as acutely as they are in smaller cities. Agriculture is a big part of our economy, and the lack of sufficient water is having a negative impact.

As a youth activist, I think sometimes the main challenge, at least in developing countries, is the lack of trust shown by experienced people, managers and authorities. Nonetheless, when we look at the climate action movement we see that the most influential people are young people. Greta Thunberg is a case in point. Despite her young age, she has started a movement that has inspired so many people around the world. Young people will confront the adverse effects of climate change, and we should have a key role in deciding how to deploy the solutions.

My parents are both proud and worried. They are proud because of my efforts to tackle climate change. But they are worried about my future, because I need a reliable income. Environmental problems are not usually given priority in developing countries and, thus, the allocated financial resources are not large. I believe that's one of the main reasons why movement towards improving the environmental situation is so slow in developing countries.

In the past, we had opportunities to secure international funding for our projects. But this path has been blocked as a result of the international sanctions on Iran. While the reason for them is unrelated to the environment or our people, sanctions affect all aspects of the country. Climate change is one of the factors that impacts directly on people's lives, and the leaders should care more about it.

The US recently renewed sanctions on Iran in 2018 that forbid US companies from trading with Iran, but also foreign companies or countries that are trading with Iran. Sanctions like this, deployed for geopolitical reasons beyond climate change, have badly affected Iran's economy.

HOWEY OU

Aged 17

CHINA

My campaign is to urge the Chinese government to fall in line with the Paris climate agreement. I started protesting because, while there are millions of people around the world demanding climate action, nobody was doing it in China.

I started by joining Greta Thunberg's Fridays For Future protest in May 2019. I stood outside my local government building demanding climate action. At first, my parents were frightened about my climate protest. They tried to dissuade me, but failed. After I was interrogated by the police, my parents stopped my protests.

To keep going, it would have taken a lot of courage. I don't consider myself especially brave. So instead, I travelled around China alone for more than two months to find like-minded people in climate and environmental protection non-governmental organisations. They understood me more.

On 13 September 2019 I launched #PlantForSurvival. We will continue to plant trees every Friday until the Chinese government is in line with the Paris Agreement. Now, I am studying in a home school in Yunnan Province, planting trees every Friday and helping my classmates become aware of the climate crisis.

China is the world's largest emitter of greenhouse gases, accounting for over a quarter of global emissions. It is the world's largest consumer of coal and the largest developer of renewable energy. The choices the country makes will have a big impact on the world's efforts to limit global warming.

THERESA ROSE SEBASTIAN

Aged 16
INDIA/IRELAND

In August 2018 I travelled to my native country of India for a wedding. Going back home is one of my favourite things to do as it is my only chance to see my extended family. But in that month the state of Kerala was hit by very heavy rainfall. This caused extreme floods in many towns across Kerala and resulted in the deaths of more than four hundred people.

My family and I were stuck in the floods. Our town, Pala, was badly affected. Outside our apartment, the water was up to my neck. My brother had to swim through it to get to town. But we were considered to be the lucky ones. Other towns were completely submerged, houses were destroyed, and people were stuck on the top of their roofs, screaming for shelter and safety. We were fortunate to find some way to travel to the airport and get on a flight back home.

Once I reached Cork, I realised that it is a huge privilege to be able to come back to Ireland and not have to start life completely afresh. (Many in Kerala rebuilt their houses, and those who could not afford to do so were left homeless. Some had to attend funerals of loved ones who had died in the disaster.) And this privilege in turn made me realise that I couldn't just sit back and wait for it to happen again.

I am protesting against the lack of action, the ignorance of reality, and the greed of profit-crazy companies. Our society has 'progressed' to the point where a piece of printed paper is more important than the lives of thousands today, millions tomorrow. The climate crisis will continue getting worse, and it will affect everyone. This is not an issue that you hear only in 'developing countries'. It is happening everywhere, from France to Angola and from Pennsylvania to India.

I have had to miss school a few times for strikes and climate conferences. I love my school, but this is something I've had to sacrifice to help make a change and get my voice heard. It is quite tough coming back and having to catch up on everything yourself. When we strike from school, we are doing it out of necessity and not out of a desire to 'bunk off'. I aspire to becoming a lawyer in the future; therefore my education is very important to me.

A lot of people tend to brush me off because of my age. They tell me that I am not mature enough to do something about climate change or even make a difference. But I have used this as more reason to make sure I am heard.

My age also hinders me in terms of attending conferences abroad. I try to the best of my ability to travel sustainably and not take aeroplanes. But I live on an island and only those aged eighteen or older can travel alone on a ferry. My parents also worry about me travelling alone to conferences, although they have been very supportive of me despite their concerns. They have had to take huge leaps of faith when it comes to my activism.

Because Ireland is a country with a small population, people consider that we can play no part in fighting the climate crisis. That's not true. I would like Ireland to become a pioneer, showing other countries how to achieve climate goals.

I am fighting for you. Join me.

On one of the strikes, I was accompanied by boys and girls who hadn't even turned ten. It struck me that these kids are also scared. Just like me, they are scared for their futures. They marched with us carrying a sign that said *I Want a Future*. After the speeches, some of them started crying. It made me realise that I cannot keep fighting just for me. I'm now determined to protest for everyone's futures, for my future kids and for future generations. I will keep protesting because I know that we still have time to fix this.

NASREEN SAYED

Aged 27
AFGHANISTAN/USA

Knowing that I come from a fragile and underdeveloped state, that I was born in a refugee camp during the height of Taliban rule, and I was the first woman in my family and village to attend university has been a source of motivation for me and inspires everything I do.

I am a dual citizen: Afghan–American. Climate change rarely gets mentioned in the context of Afghanistan. The country has witnessed severe droughts. Water and land disputes are a cause for concern. Scientists predict a decrease in rainfall and a rise in average temperatures in the coming years, leading to land degradation and desertification.

Environmental mismanagement, lack of environmental policy, trading long-term sustainability for short-term economic or political gains – these issues have prompted me to take this path. I have been exposed to questions related to sustainability and human development through my childhood. Moving from one country to the next, my interest in the global agenda, particularly environmental challenges and politics, grew. I have witnessed everything from Soviet-era chemical waste in Azerbaijan, which affected thousands of poor villagers, to drought in Sudan and to the excessive use of fossil fuels in the Gulf.

Now I live in California, and climate change plays a big role in our lives, especially with the growing number of wildfires and recurring droughts. I am worried that in the coming years I may have to evacuate my home because of the fires.

I am not a campaigner but I do participate and support many campaigns related to environmental justice and climate change. I am a member of the Citizens' Climate Lobby, campaigning for a carbon fee and dividend, and of the Sunrise movement, advocating for a Green New Deal.

LIYANA YAMIN

Aged 27

MALAYSIA/TAIWAN

I advocate for youth engagement and empowerment, especially in solving the climate crisis. I believe that young people have the power to drive solutions for combating the impacts of climate change.

Malaysia is vulnerable to climate hazards including floods and landslides, forest fires, tsunamis and cyclones. The country is currently experiencing varying rainfall, an increasing occurrence of extreme weather events, increasing sea surface temperatures and sea level rise.

In 2014 I experienced a major monsoon flood when I was studying in Terengganu and have also seen the annual transboundary haze every September/October caused by the burning of palm oil forests in Indonesia. It was an eye-opener for me that the world is facing a crisis.

With a projected global temperature rise of 1.5 °C, cities like Kuala Lumpur are expected to experience unprecedented climate conditions resulting in extreme weather events and intense droughts by 2050. People living in rural areas will also be affected by climate change. Coastal communities will be disrupted by rising sea levels. Effective adaptation measures will need to be implemented across Malaysia. This will be a significant challenge.

It is time for collective leadership on climate action instead of empty talk.

I have been actively involved with the Malaysian Youth Delegation (MYD), the only youth-led organisation in Malaysia, which focuses on climate change policy and negotiations, providing a platform for curious and interested young people to explore the world of climate agreements at the United Nations. MYD strives to educate the public on climate change policy by organising training and talks. It also maintains a relationship with the federal government and regularly engages with them.

Within the framework of our existing capitalist society, this crisis links our economy and our way of life and thus socio-environmental change is needed in every sector of people's lives: in agriculture, energy, transport, employment, and so on. We need to treat the climate crisis as an opportunity so that others don't get intimidated and will join the fight once they understand too.

During the forest fires season, winds often carry a thick smoke from Indonesia towards Singapore and Malaysia, known as the transboundary haze. Research from the 2015 haze crisis suggests the smoke may have caused 100,000 premature deaths across Indonesia, Malaysia and Singapore.

ALBRECHT ARTHUR N. AREVALO

Aged 26

PHILIPPINES

My job often involves me going to the Philippine uplands where our office runs a school for indigenous people. I will never forget my first time there. Being in the middle of an upland forest with no mobile-phone signal and being surrounded by nature for a month really changed me. My perspectives and outlooks were challenged. The area, called Bendum, used to be a logging site thirty years ago and there were barely any trees left. But with the help of partners, the community pulled together and Bendum is now one of the most forest-dense spots in the region.

The most significant moment during my first visit was when I joined a community mass one morning. I am not a Catholic and I could not understand their language, but

I was moved to tears when they sang and prayed together. The feeling of their togetherness really touched me. Despite their hardships, they remain resilient and true to their identity. Whenever I have doubts about why I continue to do the work I do, I often think of that moment.

Climate change is affecting water and food supply in the Philippines. It is also affecting the human rights and dignity of farmers, fisherfolk and indigenous peoples, who are a key part of the country's economy.

* * *

The most challenging thing about being a young activist is getting people to commit to the long-term heavy lifting that effective activism requires. It is challenging to get the commitment needed, especially living in a developing country where money is very difficult to come by. Volunteer work is not always 'free'.

Please do not forget the human face of climate change.

My parents are fine with my activism now because they are seeing the fruits. Previously, they were concerned that I was working too hard and not getting enough sleep. They were also uncomfortable with me spending my own money in order to attend events or meetings. There were several times when they asked me to stop. But I always believed in my vision and that the work I do could help many young people.

AKARI TOMITA

Aged 16

JAPAN/USA

There is a Japanese word, *mottainai*, which is best translated in English as 'wasteful'. Avoiding *mottainai* is a tradition that is a way of life for Japanese households. The idea is to express gratitude to and respect for our environment and never waste anything, especially food, energy, goods, and to always take care of your possessions and surroundings.

It is important that we do so because we are a part of nature. This includes taking reasonable portions and eating everything on your plate, turning off the light when it is not in use, and treating items with care. I was brought up with this principle, so my family was already living somewhat sustainably. However, there were still many improvements to be made at home, at school and in my community.

That is why I am campaigning for youth representation and zero waste, which is the goal that nothing goes to landfill and everything is composted or recycled instead. This also means that I advocate for less plastic use.

If I could change one thing, I would change our reliance on plastic. Almost everything is made of plastic or packaged in plastic, whether it be food, clothes, or other necessities or goods. This reliance on plastic, combined with consumerist culture, is detrimental. Plastic comes in and out of households and ends up in landfills or the natural environment, either contributing to greenhouse gas build-up or endangering wildlife.

* * *

I learned to love nature from a very young age by visiting my grandfather's rooftop garden in Tokyo and travelling to national parks in Japan and in America. I remember wanting to do something when I heard about melting ice caps and plastic pollution, among other disastrous issues, during elementary school. So I joined the elementary school's environmental club called the Green Team, and I've continued to be a member of my school's environmental clubs since then.

One of the biggest challenges to being a young activist is getting peers, especially adults but even other young people, to stop underestimating youth activists. The issue is that many people don't feel obliged to get involved and would rather humour activism or support it from a distance. The truth is that everyone needs to get involved. It's also difficult to create change at a school, because getting things through administration can be a lengthy process. But with the right people and the right mindset, change can happen.

I am nothing like Greta Thunberg or any of the other amazing youth activists who contribute so much to the fight against the climate crisis. I am not doing as much as I'd like, and certainly not as much as I should. At times, I still have trouble accepting the urgency of the climate crisis. However, I hope that I can give some advice to those who feel similarly.

With the right people and the right mindset, change can happen.

I believe the key is to stay motivated. My motivation comes from knowing that children around the world are already facing fatal health risks because of climate change. The thought of my actions hurting someone else my age is painful, and it forces me to face reality and realise that the climate crisis cannot be neglected.

Keep fighting – we cannot continue without a habitable planet.

NORTH AMERICA

CONTINENT: NORTH AMERICA

POPULATION: 580 MILLION

BIGGEST CLIMATE CHALLENGES

- **GLACIER MELT** – The Canadian Arctic's average temperature is rising at double the rate of the global average. Parts of Canada's Arctic Ocean are projected to have extensive ice-free periods during summer within a few decades.

- **AIR POLLUTION** – Higher temperatures increase the formation of ground-level ozone, a pollutant that causes lung and heart problems. In North America, more than 140 million people already live in municipalities with unsafe levels of air pollution.

- **COASTAL EROSION** – Almost all of the Caribbean's main cities, home to millions of people and their infrastructures, are less than a mile from the coast. Rising sea levels threaten millions of homes and the islands' economies, which rely heavily on coastal tourism.

- **RISING TEMPERATURES** – Temperature rises are set to increase the frequency of heatwaves and droughts, particularly in Central America and America's south-west. By 2100, extreme heat days – which generally occur once every twenty years – are projected to occur every two to three years across the US.

CECILIA LA ROSE

Aged 16

CANADA

Put yourself in my shoes. Imagine you're my age. You can't vote, and the leaders of your country aren't doing their jobs. I want you to understand how scared I am; how I worry about my family every day; and how I have debated whether or not I should choose my education or my future.

For a while now I've been attending and organising protests, calling on the Canadian federal government to protect our futures. When I'm not at protests, I work heavily in my community, and I frequently speak with politicians about the climate crisis, urging them to represent these issues in Ottawa.

Most recently I filed a lawsuit, along with fourteen other plaintiffs, against the federal government over their inaction and knowing contribution to the climate crisis. We are doing this because we are feeling the effects of climate change first hand, and we believe that our government must be held accountable. This isn't the same as my activism, because it's much more personal. This lawsuit is about our stories, and we represent young people across the country. We need the government to know that their actions are costing us, and we won't stand for it.

A big part of the reason I am where I am is because of my parents' influence. I've been attending protests and talking about politics with my family since I was really young, so I can't imagine they're all too surprised that I turned out this way. There are definitely points of disagreement when it comes to our social and political views, but I appreciate that. I've learned to open my mind more and see things from another perspective.

* * *

As a young person, I'm easy to overlook. Adults aren't used to taking direction from those who are younger than them. But this is the biggest issue facing humanity, and it is the young people who are emerging as leaders. We're not claiming to be the experts, or the scientists who should be calling the shots. All we've done is look at the facts and wonder why we're all not doing what needs to be done to protect our futures. As young people we have proven that we're more than capable of understanding the issue, and it's time that those in power listen.

It's really exhausting to constantly be told to go back to school. This is a climate emergency, which means no more business as usual. Our education and personal achievements are no longer the most important things in our lives, we have bigger goals and we're going to fight for them.

There is every reason in the world to be terrified for our future. However, that is not an excuse to do nothing. I feel that fear every day, and what's helped me the most is actually using my voice and putting my energy into bringing about meaningful change. Channel that fear into action, and get to work, speak to your friends, your family and your politicians. Demand action in every part of your life and on every level of government.

If only our political parties could argue over political opinions rather than scientific fact.

If this were hopeless, I wouldn't be here. I would go hang out with my friends and read until 3 a.m. There are many things I would rather be doing, but as long as there's hope I'll be here. The fact that young people are doing it should make others feel optimistic. We can achieve our goals. It won't be easy, but it is possible.

KAREL LISBETH MIRANDA MENDOZA

Aged 27
PANAMA

I grew up in a rural place surrounded by beautiful nature. Panama is a tropical country with great biodiversity of both flora and fauna.

But my country is also highly vulnerable to climate change, because Panama has two coasts, one on the Atlantic Ocean and one on the Pacific Ocean. The rise in sea level is affecting coastal communities, especially the Guna communities on the islands of San Blas, who are being forced to leave. Our economy is also dependent on the Panama Canal, which connects ships transiting between the two oceans. The canal relies on the water levels of two main lakes, the Alajuela and Gatun, which feed it. But climate change is affecting rainfall patterns and sometimes lowering the water levels of these lakes, threatening the safe transit of ships through the canal.

The consequences of climate change are increasing. The temperature rise in recent years has limited what I can do at certain times of the day. In the future, I might not be able to leave the house in the daytime to do regular activities because it is too hot, and there might be food and water shortages. I am afraid of such a future.

I've watched how the place I grew up in has changed beyond recognition over the years, because of bad agricultural practices, deforestation and environmental pollution.

But I don't need to tell you what's going on. Just look around and see how everything is changing, how weather patterns are changing, how nature is changing.

* * *

When I learned about climate change in college, I was terrified that nobody was doing anything about it. In Panama, few young people really care about climate change. So the biggest challenge is the lack of commitment from younger people in vulnerable countries like mine.

My campaign is aimed at making changes to the way we live, consume and discard. I'm also campaigning for local, regional and world leaders to stop harming the environment in the name of economic benefit.

Together with a group of thirty-four other young people, I formed the Youth Network Against Climate Change in Panama. I believe that the most successful method to raise awareness among young people in Panama is to mix environmental education with action and mitigation activities. Show them that they can be a part of the solution, then share it with other young people on social networks.

My mother supports me in this movement. She has even changed her lifestyle, because she is more aware of what is happening.

Leave personal interests aside and act now. Tomorrow is too late.

Participating in the 2019 UN Youth Climate Summit was a wonderful experience. I learned about what young people in other countries do, and I understood the role of communication and technology in mobilising for climate action. At this important event, young people not only proposed solutions, learned about negotiations and discussed problems; we also made our voices heard and demanded concrete solutions from world leaders. Those young people who, at such a young age, are aware that humanity is facing its worst ever crisis and who change their lifestyles and take to the streets are my inspiration.

But a grass-roots approach cannot solve the problem on its own. It is necessary for governments to commit to creating new and stricter environmental policies, and to commit to enforcing existing national and international environmental policies. Let's be the generation that saved the planet – not the one that finished destroying it.

The Panama Canal is an 82-kilometre waterway that connects the Atlantic Ocean and the Pacific Ocean, enabling 3 per cent of the world's maritime trade, and contributes more than 10 per cent to the Panamanian government's annual revenues.

EMMA-JANE BURIAN

Aged 18

CANADA

I spent most of my childhood living on Burnaby Mountain, near Vancouver, in British Columbia – on the route of the Trans Mountain Pipeline. There were many weekends when my dad would take my sister and me to the local park. We would walk through a trail that had tall grass, many flowers and buzzing bees. The trail also featured big yellow signs sticking out of the ground: *WARNING! High Pressure Petroleum Pipeline. Call Before You Dig.*

My dad explained that the pipeline carries oil. If someone dug into the ground, the pipeline could burst, and oil would contaminate the area. It really concerned my seven-year-old self. Little did I know that those fossil fuel pipelines would remain with me, even after I moved far away from those yellow signs.

* * *

Canada is warming at twice the global average. Ice and permafrost in the north are melting rapidly, and that's a concern for wildlife as well as for the people living there. The availability of fresh water has started to become a concern, and it will only get worse. Climate change is also affecting Canada's crops, because there is lots of rain in some places and very little rain in others. Dry spells can be disastrous in other ways. In 2017, British Columbia had the largest wildfire year on record. Many of my friends who live in the interior and the north had to evacuate. I was really worried about them.

I joined the Climate Strike movement because I understand the science behind climate change. I felt that, as a young person, no one was listening to my concerns. Striking was finally a tangible thing I could do to get my voice heard and make a difference. Not taking action now is simply suicide, and I am not prepared to leave a terrible legacy for future generations.

Since then, I have spent countless hours organising around the climate crisis when I should have been doing schoolwork or other things teenagers typically do. It has also meant that, unfortunately, I have spent less time with my family.

It was hard for me to learn how to handle the media, how to deal with lots of hate online, and how to balance it all with school. But the most challenging thing about being a youth activist is the feeling that you can't do it. The whole world tends to tell you that you are too young. It's hard not to believe that.

Even if we overcome that feeling and do something, it's hard to have confidence in what we have built. All of us young activists feel like we have no idea what we are doing. We were a bunch of young people who started out organising through video calls and social media. But now we are a whole organisation. We strike the first Friday of every month. On 27 September 2019 more than 20,000 people flooded the streets of Victoria to demand climate action.

Getting involved in the climate justice movement is one of the best things I've done. The community we have created is full of the most amazing people, who are truly inspiring. Being a part of this community really helps with dealing with climate anxiety and worries about the future.

* * *

One of the biggest misconceptions about young people is that we are just smaller, less capable adults. We may be small, but we are certainly not less capable. In fact, we bring valuable perspectives and talents that adults may not have. We hold an essential quality needed to bring about change. We don't subscribe to the 'business as usual' mindset.

If I could change one thing about this country, it would be the government's treatment of indigenous people. It shows the government still cares more about money than human rights. It shows why Canada is still so far behind in its climate targets. It shows how our public institutions were founded on racist and unjust ideals.

I really wish I could go back and rewrite history. Since I can't, I will put all my work into writing a beautiful future.

ANYA SASTRY

Aged 18

USA

I first got involved with activism following the 2016 US presidential election. That's because Donald Trump, the candidate who had just been elected to the highest office in the country, went against everything I believed in.

I recognised that it is not enough to simply be conscious of prevalent issues around the world. Instead, it is crucial to combat those systems of oppression and inequality by campaigning for those who have been silenced or ignored. We need to empower others and amplify their voices in order to create a more balanced and equitable community where everyone can advocate for themselves and their rights.

I only became fully aware of the climate crisis and the need to act against it when the Intergovernmental Panel on Climate Change (IPCC) published its influential 2018 report. The phrase 'twelve years left to take action' resonated not only with me but also with young people around the world as well. Seeing those words allowed us to truly understand that our future and our chance at a fulfilling life was being threatened.

* * *

The climate crisis is affecting many communities within the United States, and in a multitude of ways. Over the course of this past summer, I made a film documentary that focuses on two communities – one in Minnesota and one in Chicago – and how they are being affected by the climate crisis and environmental injustice.

In northern Minnesota, a Canadian company is constructing a pipeline through indigenous lands. The worst kind of oil, tar sands oil, will flow through this pipeline and when (not if) this pipeline leaks, biodiversity in the area will be permanently damaged and resources that the indigenous peoples rely on will be destroyed. This is an injustice, an affront to indigenous sovereignty and an environmental disaster.

The second community goes by the name of Little Village and it is located in the heart of Chicago's industrial corridor. Every day, the people of Little Village deal with high levels of pollution from fossil fuel use in power plants, factories and trucks in their local community. Kids are growing up with asthma and other health complications.

The issues these two communities are facing are also affecting many other communities across the country. And this is not the worst of the climate crisis. If elected officials do not implement immediate climate action legislation and policies, it will get worse. We will not have a liveable future.

That is why my activism focuses on preventing politicians from taking fossil fuel money, working against the further installation of fossil fuel infrastructure – especially within communities that are already marginalised due to their ethnicity, gender or socio-economic status – and campaigning for elements of the Green New Deal to be passed through Congress, working at a local level with elected officials in order to pass legislation, such as required teaching of the climate crisis in schools.

Leaders need to prioritise the lives of their citizens over power and money.

The amount of emotional and mental pressure us youth activists put ourselves under is incredible. Every day, I balance a rigorous academic course-load, intensive extra-curriculars, college applications, and relationships with family and friends just like every other student. On top of that, I make time for hours-long conference calls, press interviews, team meetings, grass-roots campaigns, and events and demonstrations. It is essentially like adding a full-time job into the mix. And what is challenging about this 'job' is that it deals with some of the most pressing issues that society is facing. To put it quite simply, we are young kids devoting so much of ourselves, our emotional well-being, and our mental capacities to combating societal issues and creating solutions in order to make our world a better place.

Thankfully, my parents are wonderful supporters of what I do. They raised me to be aware of global issues and current events, cognisant of social injustices and the wrongs of the world, and that awareness has inspired the work I do today. They also help me grow as an activist by having discussions and by challenging me to think of more effective solutions.

When I'm faced with adversity and I feel myself getting ready to give up, I remind myself of the resilience of all those I work with on a daily basis, everyone involved in the climate action movement, and those who are sacrificing a tremendous amount to dedicate their time, energy and resources to bettering the communities around them.

If you are worried about the climate crisis and are privileged enough to take action, please use your time and energy in whatever way you can. I implore you to integrate the concept of environmental justice into your actions, your solutions, your initiatives. Include it in conversations about the climate crisis, whether informal or formal. Join grass-roots mobilisations and organisations that actively combat environmental injustice within your own community or the communities around you. And most importantly, use your own platforms to amplify the voices of and empower those on the front lines, those who are dedicating their time, energy and resources to protecting themselves and their loved ones from this crisis.

The US has more than 70,000 miles of pipelines to carry crude oil. The oil is transported under tremendous pressure, which can result in faulty pipes causing a spill or a leak. Since 2010, about 9 million gallons of oil has been spilled from these pipelines.

RICARDO ANDRES PINEDA GUZMAN

Aged 22
HONDURAS

Many people don't know it, but Honduras led the Global Climate Risk Index for seventeen years as the country most affected by extreme weather events. Even though most Hondurans believe that other problems, such as poverty, security and corruption, should be the government's priority, they don't realise that the impact of climate change has already been proven to be far more destructive.

There is a dry corridor in my country that is severely affecting crops and forcing people to migrate. Heavier rainfall in other areas is causing damaging floods. But the real problem will be when the next hurricane comes. Honduras sits in between the Atlantic Ocean and the Pacific Ocean. If the next hurricane is anything like the last one, Hurricane Mitch, it will destroy everything in its path.

That is why I believe my voice is crucial. Increasing awareness about climate change is more important in Honduras than it is in any other place in the world.

It all started after picking up a book my father gave me when I was twelve years old: *An Inconvenient Truth* by Al Gore. I wasn't really into reading a thick book, but I was interested in images and graphs. Eight years later, it led me to look at what Al Gore was up to, follow him to Los Angeles, and ultimately meet him in 2019. His book sparked my interest. I thank him and pledge my life to working towards a global solution.

Young people are the most important actors in this fight. When world leaders talk about climate goals for 2030 or 2050, they are talking about a world that we will inherit. Some of us will be leaders of companies and countries. That is why we have to spread the message to other young people and empower them to face this crisis now. We have the most on the line.

Mitch was a Category 5 hurricane that hit Honduras in 1998. It became the second deadliest hurricane to hit Central America, killing more than 11,000 people. Climate change is making the oceans warmer, which is expected to make hurricanes more intense.

CRICKET GUEST

Aged 22
CANADA

For a lot of indigenous people, their culture leads them to activism. As indigenous people, a lot of the time we don't feel as though we have a choice to be activists, much of the time you're born into it. But my experience was the opposite: my activism led me back to my culture.

I'm a white-coded (or white-passing) Anishinaabekwe Métis. I was raised by my single white mom, in a tiny town of largely white people. My mother would try to take me to an indigenous education centre in my town, but when I was about four it suffered funding cuts. We were left without resources for me to learn about my culture.

But as I began to learn more about the land and the destructive effects of capitalism, colonialism and the patriarchy, it led me back to my people. It led me to the land defenders and their teachings, and suddenly I didn't feel so separate, different or radical from everyone else. Reclaiming and relearning my indigenous culture has been crucial in finding my path in climate activism.

My activism has helped me connect the dots as to why I feel so passionate about justice for women, people of colour, LGBTQ+, animals and the land. Because they're all connected. The injustices carried out against any one of those groups leads to the mistreatment of all of them.

I first started getting into activism when I was twelve years old. I began to look at the world differently and challenge systems that I didn't know were harming me up till then.

I went from being the kid that was too shy to raise my hand, even if I knew the answer, to being voted 'most likely to start a debate' by the end of high school. Later, when I was sixteen, I started learning about environmental issues, specifically how damaging colonisation and white supremacy within the agricultural industry has led to so much destruction and pain of both the land and the animals.

Use that feeling of worry and turn it into education and action.

I work with Fridays For Future Toronto, where I am the indigenous outreach coordinator. I work to ensure that indigenous voices are centralised rather than tokenised in the climate justice space. I hope that I can help make indigenous people feel safe to enter these spaces, because as indigenous people we never know what we're up against when we enter a space of large crowds of predominantly non-indigenous people.

I still see many white climate activists attacking indigenous people for speaking about indigenous sovereignty in these spaces, because they fail to see the important and vital correlation between indigenous sovereignty and climate justice. Much of the work I do attempts to bridge this gap that exists with outsiders understanding the importance of land defenders in this fight. It's not as easy as planting trees or going vegan, it's about addressing and dismantling the colonised systems that have plagued and are killing Mother Earth.

One of my most memorable experiences was when, for the first time ever, I sang an indigenous song in public at a climate rally in front of thousands of people. It was a beautiful and surreal moment for me. Learning and singing the music of my people has become therapeutic for me, but I usually keep it private. And, yet, I wanted to sing the Women's Warrior song at the climate march to honour missing and murdered indigenous women.

I still have so much to learn. I don't have all the answers. And at times, I feel as though young people are preyed on for this. I'm at a stage in my life where I need to listen a lot to grow and learn, but I am sometimes thrown into situations where I need to speak more than listen. It's a tough balance sometimes. I believe as youth activists we are strong and we are wise. But I still need to take time to learn from and listen to my elders.

My father has passed, but my mom supports my activism. Although not overtly political herself, she supports my politics and believes what I'm doing is important. She's proud of me and I'm grateful for that.

In my culture we believe that we live for the seven generations behind us, and the seven generations ahead of us, and we must act while thinking of both of them. I hope I can leave this earth better than when I arrived. It's our only option if we want to have seven generations living after us.

Métis are one of three recognised indigenous peoples in Canada, along with First Nations and Inuit. Over 500,000 people identify as Métis across the country.

LIA HAREL

Aged 19

USA

When I entered my sophomore year of high school, I met the student leaders of my school's Earth Club: a group of spunky high school seniors who laughed after every third sentence and who never looked down on a lowerclassman like me. As I kept coming back to the weekly Earth Club meetings, I grew close to them, building friendships rooted in trust and compassion.

They took me to my first youth environmental activist meeting at the offices of Climate Generation in Minneapolis. I went in expecting to hear the typical monotonous murmur about saving the polar bears, but instead I listened to discussions about climate justice, indigenous land values and lobbying tactics. What stayed with me the most was that the people who were leading these discussions were young. In their voices, I heard confidence that we could overcome the obstacles we faced. I saw in their eyes a flame of passion that we had the power to effect change. I felt the warmth of the room, the warmth of the energy of a community.

* * *

My campaign work first started on a local level. With the help of iMatter, a national organisation devoted to supporting young people in making their city governments take climate action, I decided to challenge my city, Minnetonka, to be a stronger climate leader.

Along with two other students from my school, we researched the city's renewable energy, waste and emissions reduction strategies, and carbon removal efforts. We found that Minnetonka was already taking great steps at making our community more sustainable. By moving to 100 per cent solar, the city will save 13 million dollars over twenty-five years. However, we still have an opportunity to be more aggressive in our efforts to mitigate the climate crisis. On 30 April 2018 we presented our findings to the City Council and asked them to commit to creating a climate action plan with the goals of reaching 100 per cent renewable electricity community-wide by 2030 and net zero greenhouse gas emissions community-wide by 2040.

Given that the climate crisis was not a top priority for the city, it took us over a year to make progress towards developing an official proposal for the city to consider. However, along the way we were not alone. We were joined by faith leaders, business people, non-profit workers and community members who heard our calls for action and wanted to amplify them. Together, we formed an intergenerational group called the Minnetonka Climate Initiative. We are still actively working, and we are confident that our coalition and the city can move our goals forward and make sure climate is no longer a secondary consideration.

While I have been a climate activist in Minnetonka, I have met other young leaders who are also making significant changes in their communities. In the summer of 2018 we came together to apply our knowledge from working on local-level campaigns to create a youth-led state-level campaign. We named it 'Minnesota Can't Wait', and the goal was to be an umbrella coalition that brought together many climate campaigns behind one unifying message. Our three-point platform included regulating greenhouse gas emissions, stopping the development of new fossil fuel infrastructure in the state, and passing the Minnesota Green New Deal Bill that we wrote and introduced during the 2019 legislative session.

Though our bill ultimately did not pass, there is still much to celebrate about this campaign, for we helped shift the climate conversation in the state government.

Through numerous press interviews, op-eds and rallies, we made it clear that we can no longer afford to do what is politically possible: we need to do what's necessary. We testified at committee hearings to remind legislators that it's not a Republican future or a Democrat future; it's just our future.

In all of these efforts – from local to state to national level – we brought credibility to young people's voices. We made it clear that we are not simply bodies holding up signs. We challenge political incompetence, we organise tactfully and we help bring forward the solutions. Youth has power.

A just and sustainable future requires just and sustainable organising.

WHAT WOULD YOU LIKE TO SAY TO

Elected officials are so blinded by today's profits that they cannot see tomorrow's dangers. Leaders need to prioritise the lives of their citizens. We need immediate climate action legislation. Or we will vote you out.

Anya Sastry, 18, USA

THE LEADERS OF YOUR COUNTRY?

As difficult as it might be, I would urge you to consider future generations when enacting policy. Design our system in such a way that you would be happy being born 100 years from now.

Brandon Nguyen, 20, Canada

Stop acting so small. Act as leaders, not politicians. Leaders choose courage and choose what is right over what is fast or easy.

Emma-Jane Burian, 18, Canada

SHANNON LISA

Aged 22

USA

I was born and raised in New Jersey, a state with a few infamous distinctions: the densest population; some of the highest cancer rates in the country; the most toxic waste sites on the National Priorities List. When I was about six years old, I remember being told not to play near the telephone poles near my school because the black goo (creosote) shouldn't be touched. Another year, I caught a glimpse of a 55-gallon chemical drum and men in 'space suits' on a vacant patch of land. *Do Not Eat* fishing advisory signs were dotted along too many lakes and rivers.

The toxic waste crisis in the US is far too often left out of the climate conversation, but the two are intertwined in devastating ways. Increased flooding intensity, wildfires and extreme weather have unleashed chemical wastes that might have otherwise been relatively contained and not migrated into the environment. The US Government

Accountability Office (GAO) warns that 60 per cent of some of the worst contaminated sites in the country, called 'superfund sites', are in areas that could be hit the hardest by climate-related disasters.

Hurricane Irene in 2011 and Superstorm Sandy in 2012 hit my state of New Jersey ferociously, causing widespread flooding, power outages and damage to buildings. A former American Cyanamid chemicals and pharmaceuticals manufacturing site in a town close to my home also flooded, with heavy rains causing chemical impoundments to spill over. Other sites lost power to systems that were operating to treat and prevent the spread of contaminated groundwater.

Unlike melting ice caps or beaches clogged with plastic bottles, toxic waste is often such an insidious threat because you cannot see it. Many hazardous chemicals are colourless and odourless, often holding families living near legacy sites hostage in their own homes. Corporate polluters have been able to get away with decades of self-regulation and self-reporting because the damage being done often flies under the radar of government agencies until communities start screaming or body counts start to rise.

Now my goal is that no person should have to suffer because they were born 'in the wrong zip code' where industry has left behind a noxious mess. I work to empower communities that live near leaking toxic chemical sites – many of whom are facing devastating health impacts – with the tools they need to compel the government to clean up their poisoned backyards.

I've had to learn to combine the meticulous mind of a scientist, the fact-hunting persistence of a private eye, and the gonzo tactics of an on-the-ground activist – all from scratch. This has included sealing myself in protective suits and conducting site inspections, taking scientific samples, petitioning state and federal agencies to address contaminated sites, hosting rallies and advising government agencies on their clean-up plans. As a 'toxic detective' in my late teens and early twenties, I began unravelling the chemical assaults committed against people and the environment.

For years, moms and dads in Franklin, Indiana, watched as their kids suffered and lost their life to rare, aggressive cancers. At the same time as these families were fighting devastating health afflictions, they were also fighting their government to get answers. I got in touch, and worked with the community to file extensive requests to the US Environmental Protection Agency (EPA) to get access to government documents pertaining to any known contaminated sites in the area.

After months of poring through over forty thousand pages of previously hidden documents, I uncovered a bombshell. A nearby industrial site, Amphenol, assured by the EPA to have been cleaned up, seems to have been mismanaged and poorly investigated. A cocktail of poison gases, including the known carcinogen TCE (trichloro-ethylene), may have been invading what most people consider to be a safe place – their homes – for years. The independent data we collected, and a hefty dose of advo-cacy and public pressure, led to the EPA reopening an investigation into this site.

* * *

To me, the crux of environmental activism is taking risks in pursuit of protecting human health and the environ-ment. One of the most challenging aspects of this work is going up against entrenched government agencies and multinational polluters, who outmatch you in money, political contacts, and scientific and legal expertise. This is only further magnified when you're a young person and are an easy target for detractors, who attempt to shut you down because of your perceived lack of technical knowledge.

But while this can be daunting, I keep in mind that when I go home, I do not have to face the anxiety of poison gases coming up into my home. On-the-ground individuals in their own communities struggling with pollution, particularly marginalised groups and indigen-ous people, are the real day-in, day-out heroes fighting for the survival of all of us. It is only my hope that I get to play a role in this narrative, get to inspire others to their own podiums, so that no family has to feel unsafe in their own home – in their own country – because of industry's toxic assaults.

KHADIJA USHER

Aged 26

BELIZE

I come from Belize, a very small developing economy. With a little over 370,000 people, my nation has never had the resources to advance at the technological pace of the developed world. I have always observed that our delays in development have placed us in compromising positions. Whether it be transport, medicine, energy, agriculture or tourism, our sectors have been subjected to outdated solutions, primarily because we can't afford better ones.

My country is home to the second largest barrier reef in the world, and we take the preservation of our maritime ecosystem ever so seriously. This became evident to the international community when, in June 2018, UNESCO revealed that Belize's Barrier Reef Reserve System had succeeded in escaping the list of endangered world heritage sites due to over a decade of work to protect this maritime ecosystem that we hold so dear.

It is my belief, however, that advocacy, in itself, was and continues to be only one of the many associated tasks in our role of addressing the climate crisis. What the narratives fail to address is the role of young people in bridging two generations: a generation committed to economic development, and a generation advocating for environmental preservation. Without bringing these two perspectives together, we will not have succeeded.

Young people must be able to demonstrate to the economists that a sustainability movement will not deter national productivity; rather, if instituted properly, it will enhance it. Similarly, young people must be able to translate the concerns of environmentalists, and in doing so, seek generational harmony in this limbo time we find ourselves living in.

BRANDON NGUYEN

Aged 20

CANADA

I'm campaigning for accessible climate education. It's important that everyone has access to adequate information and the resources to properly understand how their actions will affect the environment around them. A March 2019 poll showed that the overwhelming majority of parents in the US, regardless of political affiliation, want their kids to learn about climate change in school. But less than half the parents and teachers actually teach and talk to students about climate change.

This discrepancy is driven largely by the fact that many teachers feel like climate change is outside their subject areas, or they don't feel like they know enough about climate change to be able to teach it. Through my work with various non-governmental organisations and advocacy organisations, I'm hoping to develop more accessible climate education materials that intersect various different subject areas. Climate change shouldn't be an isolated topic of interest, but an integrated and cross-disciplinary lens from which to study and understand other aspects of the world.

* * *

I'm very privileged to live in a place where the immediate effects of climate change aren't yet being felt on a daily basis. But that's not the reality for many people in the world.

Climate change produces a sort of existential anxiety in me. I feel an overwhelming urge to change the course of society. We've heard from so many scientific reports that we are rapidly reaching the point of no return, and I can't help but feel an overwhelming desire to use my resources and privilege to advocate for systemic change for those who can't advocate for themselves.

Complacency can be both dangerous and challenging to an aspiring change-maker. As a young person, there always seem to be a million things that I should or could be doing outside of environmental organising and activism. By its very definition, activism is calling for change and challenging the status quo, which can be exhausting at times.

It's an art to find a balance between taking care of myself, not neglecting the typical responsibilities of a young person – doing well in school, working various jobs, spending time with friends – and thinking critically about the world and how I might be able to change it for the better. And while I'm nowhere near mastering this art, I've definitely gotten better through failing and learning from it.

My parents have always been super supportive of my activism and organising, and I am so lucky and grateful for that. While they still urge me, every once in a while, to try and find a place in the world that balances passion with practicality, they are mostly just concerned about me burning out because of the exhaustion that comes with constantly trying to bring about change under a system that isn't conducive to change.

VIVIANNE ROC

Aged 22

HAITI

I live in a ghetto. When it rains, the streets are often flooded and sometimes water enters our homes. Haiti is a small country, one of the poorest in the world. Our people live in excruciating misery, and now comes another problem: climate change, which has turned everything upside down.

It is much warmer now in Haiti. There are many more mosquitoes and therefore more vector-borne and water-borne diseases. That means I am more likely to get dengue or malaria. Our seasons are also getting weird, causing a scarcity of food, which drives up prices and increases rates of famine. Natural disasters have affected our economy, and we have lost many who were dear to us.

Our country is suffering from the effects of climate change, and we have had enough. I am campaigning for real and lasting change because, as a young woman living in a poor country, I am affected by climate change. I want my voice and that of other young people like me to be heard.

Six years ago, I had no idea about climate change. I lived my life as a quiet teenager, not worrying about the climate, just like many young people today. An organisation that advocates for the cause is the Caribbean Youth Environment Network (CYEN). It has helped me enormously to become the person I am today. CYEN helped me realise the danger our planet is facing.

Now I chair Plurielles, an organisation which fights against the ravages of climate change and promotes the involvement of young people, especially women. They must become aware of the problem, and then they must, like me, take their place in this struggle. Education is the key to success. When I face obstacles or when I feel like I'm going to crack, I think about the next generations and I gain the strength I need to continue.

Don't be a spectator, act instead.

Major economic powers must stop pretending they are doing everything they can because it is not true. We are trying to survive, but if nothing changes in the next few years and if climate targets are not met, it is very likely that Haiti will disappear under the water or be destroyed by disasters. My future is in danger.

OCTAVIA SHAY MUÑOZ-BARTON

Aged 16

USA

I'm blessed to have grown up surrounded by nature. I was born on the Californian coast, and I spent my childhood falling in love with the Pacific Ocean and our local ecosystems. When I was younger, I was vaguely aware that humans were responsible for the pollution I saw, but I had no idea of the extent to which humans have devastated our environment and threatened the stability of the biosphere.

When I was about twelve, some kids from an organisation called Heirs To Our Oceans gave a presentation about their work to my class. They offered to take us on a trip kayaking in Elkhorn Slough, a nearby estuary with a thriving sea otter population. While we were out on the water, we were taught about sea otters and the threats they face.

Then, in 2019, I took part in the Summit for Empowerment, Action & Leadership (SEAL), a two-week youth event organised by Heirs To Our Oceans that brings together young people from all over the world to connect and learn about our environment. During SEAL, I began to understand what it meant to be empowered, and that changed me forever.

At the end of the summit, all the international youth, from Micronesia to Africa to New Zealand to Kentucky, united by a commitment to protecting our environment, came together for a final presentation. After two weeks of public speaking workshops, connecting deeply with our environment and ourselves, and training to be empathetic leaders, I watched friends stand up in front of their communities and speak some of the most powerful words I've ever heard.

I watched my friend get up and recite a poem, straight from the heart, that I would never have imagined them being able to share. Power, love, strength and confidence were the defining energies on stage. When I got up to speak and perform, I was close to crying. I felt braver and safer than I had ever felt before. I knew I now had a family of kids who were going to change the world, who would grow up to become true leaders.

* * *

While I've had an amazing experience in activism, I've had days when I feel hopeless. Being frustrated by people who won't listen to my concerns, watching world leaders draw our attention to irrelevant issues while they take away our clean air and water, and internalising the realities of what we are facing can be devastating. Sometimes, I feel like I'm losing my childhood.

I have found myself losing the motivation to study and prepare for the future. When my mind is consumed with thoughts of what life for my kids might be like on a planet ravaged by climate change, it's hard for me to focus on my assignments when I know I could be out there fighting.

My parents are incredibly supportive. Honestly, I'm astoundingly fortunate. I know some young people whose parents are ambivalent or even discouraging towards activism, and having parental encouragement makes a huge positive difference for kids interested in pursuing advocacy.

That said, my parents raised me to care about my education, and they've been understandably bothered at the thought of me not fulfilling my academic potential. I'm working towards developing a system in school where youth activists get academic credit for their work (research papers, speeches and hands-on citizen science relevant to real-world issues are hugely educational and should be recognised as such) and I'm working hard to keep my grades up.

To kids like me, the climate crisis can be terrifying. The important thing is not to let that fear paralyse you. Use that fear as motivation instead.

Surround yourself with people who empower you.

PAYTON MITCHELL

Aged 21

CANADA

I grew up within sight of the Nanticoke power plant, the largest coal-fired power plant in North America. Because of the power plant, I had chronic asthma throughout my childhood, which finally subsided when the plant closed down in 2013.

Canada is actively perpetuating climate change through the expansion of fossil fuel projects like the Frontier tar sands mine, the Trans Mountain Pipeline, the Énergie Saguenay pipeline, and the Alton Gas storage facility. Each of these projects is a direct attack on indigenous people and their land.

The impacts being felt include an ongoing First Nations water crisis and the loss of culture and traditional ways of living. Fossil fuel extraction projects and other polluters have made the water in many First Nations communities undrinkable. Melting snow and ice in the

Arctic increases the risk for northern residents and causes the loss of culture for the Inuit, who have built their entire way of life around ice and snow.

Canada's reliance on oil and gas has contributed to the rise of petrol nationalism, a far-right movement which regards Canada's oil and gas industry as our national identity and accuses foreigners of trying to kill the sector. Camps for workmen which are erected along pipeline projects contribute to the abuse of indigenous women who live on the land.

* * *

I've seen the impacts of climate change through increased floods and shorter winters. I've seen it in tornadoes in places that never used to get them. But I'm fortunate enough not to be too directly impacted from the climate crisis, yet.

But climate anxiety affects almost every decision I make. I get anxious and will suddenly burst into tears over the state of planet Earth. I get into fights with friends and family who tell me to care less or to focus on my school. It's depressing and devastating, and I know I have it easy compared to most, which makes me all the more sad.

Luckily, I am able to turn these emotions into actions. I truly believe that the youth climate movement is the first step in building a global family that, instead of being driven by competition, is driven by compassion for others, the earth and ourselves.

No one pays us for what we're doing. Despite many non-governmental organisations benefiting from the climate movement and using our work to ask for donations, we the strikers have to sacrifice being able to graduate on time, work part-time jobs, and lead regular lives in order to make all this happen.

My parents are a mixture of proud and concerned. They're proud of me for overcoming obstacles and fighting for what I believe in, but they're also worried I'm not taking my future seriously. They're concerned about how striking affects my ability to graduate, and about how much time I spend working on the climate movement as opposed to working jobs to support myself.

I've had to sacrifice time with friends and family, time from school (which I pay a great deal to attend) and time to simply enjoy being young. I am proud of this movement, but there is no denying that it has stolen and will continue to steal the best years of our lives. It's worth it, but it's not a sacrifice we should have had to make.

Despite wanting to be a mother, I've decided I can't unless this crisis is averted. Not for population control but because I don't feel safe raising a family in a world that has hit the self-destruct button. After reading about the 2018 Intergovernmental Panel on Climate Change (IPCC) report, I realised that I wouldn't be able to have a family if the world continued on this path of maximising profit and minimising accountability.

If I could change one thing, I would have Canada cancel all existing and planned fossil fuel extraction and transportation projects while committing to a social safety net to protect workers as we transition to renewable energy.

If you don't think you can do it, know that you can.

ASHLEY TORRES

Aged 23

CANADA

Student Ashley Torres gave the following speech on 13 July 2019 in Montreal at an Extinction Rebellion sit-in protest, where twenty-five activists were later arrested.

Dr Martin Luther King once said our lives begin to end the day we become silent about things that matter.

We live in an age where our values are being challenged.

A time where unjust government bills are telling our Muslim sisters that wearing their hijab will cost them their job.

A time where our neighbour, the US, is separating children away from their asylum-seeking parents, because the leaders do not understand how desperate the parents must have been to take their kids through such a horrible journey.

A time where indigenous women are going missing and no one in power seems to really care.

A time where powerful men keep getting away with sexual harassment because their money and their power protect them.

A time where our irresponsible provincial government and American companies such as Freestone and Breyer Capital are trying to build the GNL pipeline while we are in the climate crisis.

A time where our federal government declares a climate emergency and then the next day approves the expansion of the Trans Mountain Pipeline that will make it easier to extract the dirtiest oil on the planet.

A time where our environment minister brags about our clean water, while only a few miles away our Attawapiskat brothers and sisters do not have access to clean and safe water.

A time where our national security is not being threatened by a distant foreign entity but by a climate crisis that our government refuses to protect its citizens from.

A time where young people have no other choice but to miss school every Friday because they feel that no one else cares about their future.

Now, some of you are probably confused about why I have chosen to talk about topics that are not directly connected to the environment. Well, all of this is connected. When we talk about climate justice and we talk about not leaving anyone behind, we have to talk about issues to do with race, gender and class.

We will all have to face the consequences of climate change at some point, but most of our marginalised communities have been facing them already. So if we truly want climate justice, they must always be on our mind when we talk about solutions. While some of us are here today to fight for our future, we have to remember that we are also fighting for their present.

We are at a time when we must decide what changes we want to see and what we are willing to do to see them come true. The only thing I know for sure going forward is that society has never transformed by itself. It was changed by individuals like you and me who are ready in a peaceful manner to put everything on the line for a cause that matters. Although this fight sometimes seems too hard, we have to remember our lives begin to end the day we become silent about things that matter.

SOUTH
AMERICA

CONTINENT: SOUTH AMERICA

POPULATION: 420 MILLION

BIGGEST CLIMATE CHALLENGES

- **AMAZON RAINFOREST** – Comprising billions of trees, the Amazon spans nine countries in South America and is the largest rainforest in the world. The Amazon absorbs a quarter of the total carbon that is taken up by forests around the world every year. But continuing deforestation puts the forest's ability to continue to be a carbon sink under threat.

- **GLACIER MELT** – More than 99 per cent of the world's tropical glaciers are in South America, with over 70 per cent found in Peru, and they are crucial sources of water for many countries in the region. However, warming temperatures have caused glaciers to rapidly recede – 98 per cent of Andean glaciers have shrunk this century.

- **EXTREME TEMPERATURES** – In a world that's warmed up by an average of 4 °C, 90 per cent of South America will experience heat events that are currently only experienced every 700 years.

EYAL WEINTRAUB

Aged 20
ARGENTINA

My climate advocacy began in February 2019. I was on social media when I saw that people were sharing videos of Greta Thunberg. At the time I did not yet know who she was, but I saw that she was calling for an international climate strike on March 15th.

I started to investigate and found out that no one was actively organising the protest in Argentina. So I gathered a group of people and we founded Jóvenes Por El Clima Argentina (JOCA), which stands for Youth for Climate Argentina.

Our first goal was for Argentina to participate in the international climate strike. Our hopes weren't high. We had less than a month to prepare, and had no experience or resources.

I woke up at 6 a.m. on March 15th to make sure everything was ready. Expecting a few hundred people to turn up, we had prepared a sound system and one microphone to ensure some people could give speeches. By around 5:30 p.m. more than five thousand people had decided to join us to demand climate justice in front of the Palace of the Argentine National Congress.

It was this moment that convinced me of the power of grass-roots movements, that change doesn't always come from the top. It comes when millions of people get together and demand it on the streets.

Since then, JOCA has become the biggest youth climate movement in Argentina. We don't just organise strikes, we are pushing the government to adopt laws that address the climate crisis. In July 2019, Argentina became the fourth country in the world and the first in Latin America to have its lawmakers declare a climate and environmental emergency.

We did not achieve all this by ourselves. One of the main reasons for our success is that we understand that social advocacy must be intersectional. You cannot fight for climate justice without also fighting against other injustices: racial, gender or economic. This has allowed us to work together with many other organisations, amplifying everyone's voices.

* * *

My protest is against a system that prioritises the interests and greed of a few against the needs of the many. A system in which twenty-six people own more wealth than the poorest 3.8 billion people combined. I do not believe that in a capitalist system we can provide the solutions needed to avert the deepening of this disaster.

Currently climate change does not affect me and my standard of living in any significant way. But there is a very specific reason for that. I am a white, middle-class male. I am extremely privileged. Climate change affects the socially vulnerable in greater numbers. It hurts most those who have contributed the least to global warming. Climate justice is not only about mitigating the amount of CO_2 emitted to prevent future damage, but also about coming up with adaptive measures for those affected.

Worrying is only useful if we use it as a catalyst for change. From my perspective, the most important thing to understand is that individual actions will not save us. Recycling is great, going vegan is amazing, as is cycling more and lowering the amount of fuel we use. But the only way we will be able to achieve the mass transformation necessary to avert the worst impacts of the climate crisis and reverse the damage already done is if each and every individual who worries decides to get involved.

The only way to achieve systemic change is with mass movements, and mass movements need millions of people working together, not just individuals who change their personal habits.

Think globally, act locally.

DANIELA TORRES PEREZ

Aged 18
PERU/UK

Part of my family still lives in Peru. Some on my grand-ma's side are members of a tribe in the Amazon forest. Visiting Peru and seeing things worsen so rapidly in recent years has opened my eyes to climate change.

Peru has been badly affected. In the past few years there's been an increase in the frequency and intensity of severe weather events such as floods, droughts and rain-storms. In the future, it will only get worse. The rising sea levels could make some popular areas on Peru's coast uninhabitable in eighty years. Peru's glaciers are melting, and further floods and droughts will gravely affect crops, contributing to food scarcity.

Climate anxiety is a real struggle. Feeling powerless as an individual is overwhelming. Not knowing whether my family in Peru will have to be evacuated from their homes again, or if they will become victims of climate change, is terrifying. My future is uncertain because of climate change, not to mention that of future generations.

We are nothing without hope.

Initially, my mother was against my protesting. But in 2019 she finally understood the importance of climate change and gave me her support. The first protest that I helped organise was in February 2019. We expected a low attendance because we had just started organising. We had a turnout of five thousand students. I cannot explain in words how beautiful it was to know that people care. It restored my hope in humanity.

> Peru's glaciers are the main source of the country's drinking water. Peruvian glaciers have shrunk by 30 per cent since 2000. The loss of ice is increasing rapidly, putting them at greater risk of disappearing in coming years.

CATARINA LORENZO

Aged 13

BRAZIL

I have grown up close to beautiful coral reefs found near Salvador in Brazil. My mother and I have swum there all my life. In the summer of 2019, when I got close to the big coral reef, I saw a lot of white dots on it. These dots indicate that the coral is dead.

I thought the reef had died because the upper layer of water is too hot. So I dived and touched the sand at the bottom. Even the water there was hot. I couldn't stay in the water for very long. If I can't handle the heat, how can fish, coral reefs and other sea creatures?

Don't be scared. Together we are stronger.

I didn't know what made the water so hot. Then at school we were taught about climate change, and how human activity is causing temperatures to rise. From that day on, I knew that I had to stop climate change. The experience of seeing dead corals is like a hand on my back pushing me to fight.

* * *

Here in Brazil we are suffering more droughts than in previous years. Sometimes we have less rain and sometimes we have rain in excess. In the past, people would say 'This is the date of the rain' and they would prepare their crops, and it would rain. But now the climate is so crazy that people can't say this any more. People have lost a lot of food and money.

In my city, when it rains heavily, the sewage is dumped in the river which goes into the ocean. I'm a surfer, and sometimes I can't even enter the water for fear of getting sick. It doesn't just affect me, it affects sea life too.

There have also been many more fires in Brazil. Everyone saw the burning of the Amazon, but another fire which concerned me more was in the Pantanal wetlands. This is a humid place, which should have a lot of water, not fire.

Climate change is taking away my right, and the rights of other kids and young people, to have a future. Our leaders need to listen to what young people are saying because we are just trying to fix our house, planet Earth.

The Pantanal wetlands is the largest tropical wetland and one of the most biodiverse regions in the world. In 2019 over eight thousand fires were recorded in the area, up 462 per cent on the same period the previous year.

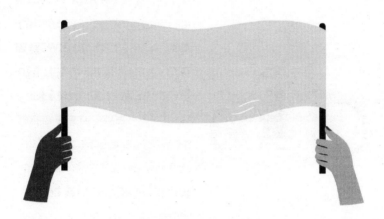

My most influential political inspiration is Bernie Sanders. He has been an activist his whole life. He understands that people in power will never lose any of their privileges out of the kindness of their hearts, and that the only way to achieve real change is to start from the bottom and reach the top by getting people involved and inspired.

Eyal Weintraub, 20, Argentina

INSPIRATIONS?

My biggest inspirations are my family, who are always telling me how important the environment is, and who are always at my side helping me in this fight; Greta Thunberg, because I see what she is doing and it makes me want to do more; and also everyone who tries to do even just little actions that help.

Catarina Lorenzo, 13, Brazil

Lots of people inspire me, but the ultimate answer to this question is nature. There's nothing more gorgeous than nature itself and its different ecosystems and forms of life. Thinking about it gives me the much needed energy to keep going down this path.

Juan José Martín-Bravo, 24, Chile

JUAN JOSÉ MARTÍN-BRAVO

Aged 24

CHILE

In school, watching the documentary *An Inconvenient Truth* by Al Gore changed everything for me. From then on I decided I had to change what was happening and that I would dedicate my professional career to it.

Chile has seven of the nine characteristics of vulnerability to climate change, as defined by the United Nations: low coastal areas, arid and semi-arid areas, forest areas, territories susceptible to natural disasters, areas vulnerable to droughts and desertification, urban areas with atmospheric pollution and mountainous ecosystems.

For example, most Chilean cities get their water from the mountains' snow and glaciers. Both sources are highly dependent on the stability of the water cycle, which is affected by rains and global warming.

The climate crisis is not just about us – it is an ecosystem crisis. Chile has experienced more than a decade of drought. Flora, fauna and fungi have disappeared from my city during my lifetime. Wildfires have consumed the forests that I knew, and the mountains that surround Santiago have succumbed to desertification. When we say 'leave no one behind' we must mean it and recognise that the most vulnerable include many humans, and many more non-humans, and we must fight for them all.

* * *

For the past six years I've been working on sustainability and fighting against climate change. It's been a long road of growth, learning, dreams and hope, and what I campaign for has evolved. It started with promoting renewable energy when I was studying engineering. Now it's about building an organisation that provides a space for people to unite on climate action. My goal is to help scale sustainability solutions, break stereotypes and build high-impact projects.

Of all the projects I've worked on, there's a special place in my heart for the Operaciones Cverde project, which went on to win the National Environment Award, dubbed Chile's Nobel Prize for environmental work. We started with no previous experience and dedicated ourselves to learning from other people. It was a volunteer project that was to help manage the restoration of wetlands in Pichicuy and El Trapiche. The most wonderful part of the work was the unintended benefits, such as birds coming back to the wetlands to make their homes and the friendships with locals who worked with us.

We live in an era of information, which requires both young and older people to exercise humility. Older people should accept that information is easily accessible and that young people have access to it. But young people should also accept that no amount of information can deliver what experience provides. We need to work together: if we have the information, let's prove it; if we have the experience, let's share it; and if we have done the work, let's show it.

Now is the time for the environment and people to be the first priority.

JOÃO HENRIQUE ALVES CERQUEIRA

Aged 27

BRAZIL

I have a project where we travel by bike to hear the stories of people at the forefront of the climate crisis – especially in traditional and indigenous communities. I've learnt that the last ten years have been the worst for droughts and storms, and many people are being forced out of their territories by climate stress.

In Brazil, we have more cases of drought than ever, alternating with severe storms. This puts the most vulnerable communities at risk every day, both those in rural areas and those in urban slums. The largest cities in the country are in coastal areas vulnerable to sea level rise. The Amazon rainforest could be degraded to the point of losing its original characteristics, becoming a large savannah.

I campaign because I feel that as a society we are heading for total collapse and that most people are not aware of it. Many people do not believe that activism is necessary or that the climate crisis is relevant. Although we do many things with little resources, it would be amazing to have some financial sustainability so that we could focus more on activism.

I am inspired by all the warriors who gave their lives defending their territories and nature – something that is very common in Brazil. My parents worry about the exposure I get. They are afraid that I might be threatened or subjected to some kind of violence.

Land and income is concentrated in the traditional sectors of society in Brazil. This has created a lot of inequality since colonial times. These sectors have a lot of economic and political power, and could help the country follow a less extractive economy. We are working to ensure that we can elect representatives who believe in science and work to strengthen democracy.

My message to people who are worried about the climate crisis is this: take action. We have no more time to wait for the transformation we need, and we cannot continue to believe in infinite development. We need to change our consumption and elect decision-makers aligned with the environmental agenda. Not to do so now is to invite the worst consequences of the climate crisis.

GILBERTO CYRIL MORISHAW

Aged 25
CURAÇAO/THE NETHERLANDS

I was born on an island called Curaçao in the Caribbean. Sunny skies and blue water were a part of my childhood. But so was a polluting refinery, which can be found in the middle of Curaçao's harbour.

I remember that my high school would sometimes be shut because the wind had changed direction and that would have brought toxic chemicals from the refinery. That wasn't the case with all the schools that were down-wind from the refinery; some of those students breathed in the noxious fumes.

While I understood the harms of toxic pollution, until recently I always took a stable climate for granted. The temperature in the tropics was the same all year round, which made me forget about seasons. But now I know that climate change is degrading Curaçao's beautiful coral reefs and affecting our coasts. Those reefs were once protective barriers against storms. Climate change is also affecting fisheries, which form a key part of Curaçao's economy.

Curaçao is a dry semi-arid island. Under most scenarios, climate change will make life worse for my people: more droughts, less rain and more heat. As the sea level rises, Curaçao runs the risk of flooding and the island will become less habitable. Poverty and food insecurity will increase and people will be displaced.

I have understood that climate change is symptomatic of a bigger issue. It's really about our relationship with ourselves, our relationship with others, and our relationship with the world around us.

In that sense, all injustices are interconnected. When we look through history and we see many different examples of crises, injustice and oppression, we can also see the similarities between them. I started working on climate change because it is also connected to fighting for an inclusive society, fighting for economic justice and for the way we treat people.

People on Curaçao need to know that climate change isn't something that is just happening abroad; it is also highly relevant to us. We need to be honest about how we have contributed to climate change. Our local refinery is one of the most polluting refineries in the world. Until recently, we were one of the highest per capita carbon emitters in the world.

We can and must do better. Curaçao has the potential to be an example for how small island developing states can take the lead on climate action. We need to start innovating our systems and understanding that change comes with discomfort.

At the same time, we must do all in our power to ensure a just transition for those who are most vulnerable. We must provide them with the power and opportunity to create a life that is unharmed by unjust fossil fuel operations, which have been poisoning them in the name of financial stability and economic independence. We can do it, we will do it, and we will win.

We cannot allow fear to steal our courage.

EUROPE

CONTINENT: EUROPE

POPULATION: 740 MILLION

BIGGEST CLIMATE CHALLENGES

- **HEATWAVES** – In 2019, the UK, Belgium, Germany and the Netherlands all saw record temperatures – and the hot weather experienced in the Netherlands and France was made 100 times more likely by climate change.

- **CLIMATE MIGRATION** – The number of people trying to cross the Mediterranean Sea and seek asylum in Europe is rising as global temperatures increase. Droughts and lower crop yields will proobably increase, driving more people from the worst affected areas to attempt the risky migration route in greater numbers in the decades to come.

- **WILDFIRES** – Between 2008 and 2018, an average of 464 wildfires were recorded every year. In 2019, this number more than tripled to 1,600 and over 270,000 hectares were burned. This area could increase by 40 per cent under a 1.5 °C temperature increase, or 100 per cent under the worst-case scenario.

- **INCREASED FLOODING AND STORMS** – Northern Europe is likely to be subject to heavier rainfall and more likely to experience flooding.

HOLLY GILLIBRAND

Aged 15
SCOTLAND

Holly Gillibrand lives in a small town outside Fort William, Scotland, in the remote, mountainous Scottish Highlands. In April 2019, she gave the following speech at the Scottish Greens Conference.

My name is Holly and I'm a thirteen-year-old conservationist and environmental activist from Fort William. I recently became a young ambassador for Scotland the Big Picture and I campaign against wildlife persecution for the animal protection charity OneKind. I have also been school striking for the climate every Friday since January 11th.

Scientists estimate that every single day 200 species are lost forever. Humans have caused the extinction of over half of life on Earth since 1970 even though we as a species have existed on this planet for a minuscule amount of time. In Scotland, one in eleven species is at risk of extinction, and Britain was ranked number 189 out of 218 countries that were assessed for biodiversity intactness. Out of 218 countries, only 28 are in a worse ecological state than us. I started striking from school not so much for humanity but for all the animals that we are dragging off the cliff with us.

To quote Greta Thunberg, 'We have done our homework and our leaders haven't.' Westminster is so entangled in the Brexit mess and obsessed over economic growth that the politicians have no thought for anything else, not even the survival of life on this planet.

Although Scotland claims to be a world leader for action on climate change, we are in the top twenty countries for carbon emissions in the world. If everyone lived like people in the UK, we would need the resources of 2.9 planets to support us every year. The youth wants change and we will not sit in silence while the adults in charge exploit and pillage our planet's life support systems and continue to encourage exploration for oil and gas.

The oceans are acidifying, rising and warming. The rainforests are being cut down. The polar ice caps are melting. Coral reefs are being bleached. Our seas are overfished and full of plastic. Fields are sprayed with pesticides. Extreme weather events are becoming increasingly common. And the extinction rate is 1,000 to 10,000 times higher than what is thought to be natural.

This is the world that young people my age will have to deal with once the older generations are no longer here. And this is the world that my generation will not accept. Adults congratulate us and say it is the youth who will save the world. It is not the youth who will save the world. It is all of you, the grown-ups, the adults, the politicians, who must lead the fight for our planet, for our futures and for all life on Earth. We don't have time to wait until we grow up.

Much of Scotland is a barren ecological desert, devoid of wildlife. Up to 20 per cent of this country is used for grouse shooting, an outdated blood sport where a large portion of land is used for the enjoyment of a very small number of people during a short period of four months each year. Wild animals native to Scotland are shot, poisoned and trapped by gamekeepers to protect the red grouse, which will then be shot too.

This has to change just as our addiction to fossil fuels has to change. The protection and restoration of the natural world is essential to mitigate the destruction that climate breakdown will cause. Scientists talk about future technologies that will capture carbon dioxide from the atmosphere, but these technologies already exist. They are called trees and peatland. These are the natural solutions that we must not forget. It is up to us to make our leaders clear up this mess, and we will not stop fighting until they start behaving like the adults we once believed them to be. Thank you.

Don't congratulate young activists, listen to what we have to say.

STAMATIS PSAROUDAKIS

Aged 22

GREECE

I am campaigning for a future. Not a better future, but a future of any kind. Humanity as a whole is at risk.

I am protesting against capitalism, greenwashing policies and profit-driven ideologies, and against blindness, apathy and ignorance – which together distance citizens from exposure to the climate catastrophe. I am campaigning for the right to live.

I started being a voice for environmental justice once I learned about the negative impact and consequences that daily choices – where we shop from, what food we consume, who makes our products – have on the world's climate. I was shocked to discover the hidden environmental cost of meat and fast fashion. I am not satisfied by the policies and laws on combating climate change. I believe that the newly introduced measures are mostly greenwashing policies that do not solve the problem as a whole.

The most challenging part of being a young activist is the sheer effort of constantly having to educate and prove the validity of our arguments to older policymakers. Because of ageism, young people are mislabelled and often not taken seriously. At a time when young people globally are directly affected by the climate emergency, we should aim for cross-generational cooperation, not judgement and an endless 'blame game'.

I am inspired by anybody who understands what the climate emergency really is and who has decided to act. I am especially inspired by activists in places where the environmental debate doesn't get attention.

* * *

Greece is facing a big threat when it comes to the climate crisis because many of the country's islands are highly vulnerable. We have been experiencing extreme weather conditions in the country over the past few years, but few people talk about climate change. Currently, I am breath-ing toxic air every day, living in too hot a climate, and worrying that the next extreme weather event will destroy my home. I am worried that climate change will radically affect my life at any moment.

I am a big advocate for the most vulnerable voices being on the front line of the debate. Greece, along with other countries that have vulnerable coasts, should be a catalyst and driver of the environmental movement in Europe.

It is clear to me that the privileged countries who are responsible for the climate catastrophe are ignoring the issue because they have built a safety net which protects them from the consequences. The ones who suffer the most are in regions that lack the infrastructure to deal with climate change. Such countries have experienced the most brutal impact of environmental reality, seeing homes being destroyed, people drowned, and many fleeing their country and becoming 'environmental refugees'.

Actions speak louder than words.

LILITH ELECTRA PLATT

Aged 11

THE NETHERLANDS

I protest against the lack of responsible government action to slow down and prevent the effects of climate change and plastic pollution on the natural and living environment. I campaign for a safe and happy future for the children of the world; for living in harmony with nature and respect for all living creatures; and for an end to all social discrimination.

I began in 2015, when I was seven years old, with cleaning plastic and other rubbish from the streets and public facilities of my home area. That developed into Lilly's Plastic Pickup project: collecting, sorting and recycling rubbish regularly; and publicising the results, messaging others to get involved and encouraging others to follow.

I was inspired to protest against the lack of action on climate change in September 2018 when I saw the pictures of Greta Thunberg's school strike actions in Stockholm. I said 'I have to do this' and began a weekly protest action outside my local town hall.

The Netherlands is not as badly affected as many other countries, but with more than half of the country at or below sea level, it is very vulnerable to rising sea levels.

The most challenging thing is coping with the negativity, the personal insults and the anonymous social media trolling. There are people who come to my school strikes and shout that it's stupid and that we should go back to school. They have lost their green heart. The green heart is a human's bond with nature. Politicians and grown-ups need to find their green heart again and see why it's so important that we have this bond and that this will help save the planet.

I go forward on the basis that my cause is just, that the future must be secured for the young, and that in the end we will win. We keep on, we are not disheartened, we are many and strong together, and our motto is, 'If we build it, they will come.'

ANNA TAYLOR

Aged 19

ENGLAND

From a very young age, I felt invisible. For so long the natural environment had been immensely important to me, but no one else my age seemed to be aware of the threat to it. So I wanted to create a platform from which other young people like me could work together to raise awareness about climate change, and make it easier for more young people to become engaged in it.

At first I started campaigning within my school and attending marches led by other organisations. I then founded the UK Student Climate Network and coordinated the #youthstrike4climate movement in the UK, which involved a huge amount of work.

A memorable moment of my activism was when I spoke to a fellow activist in the Pacific Islands over Skype. Although I'd been campaigning for months, it's still so easy to feel distant from the harsh reality that catastrophic effects of climate change are happening right now in certain parts of the world. Speaking to her was a reality check, and hearing about the extreme weather events her community has to deal with on a frequent basis brought me to tears. I'm only sorry I can't do more. And ever since that conversation I have been motivated to campaign for climate justice on a much deeper level.

At the start, my parents weren't that supportive. But I think they don't mind it so much now. It's immensely hard being a youth activist at the same time as maintaining your emotional well-being. Youth activism has given me a sense of purpose, motivation, empowerment and community, and has benefited my mental health in so many ways, but too much of a good thing can be bad. For some time, youth activism also consumed my life, and it was the only thing that mattered. I was working 24/7 for what seemed to be very slow progress.

Although I just find it amusing now, at the start the hate I received on Twitter really surprised me and was very deflating. And having to deal with all of this on top of doing my schoolwork was completely overwhelming.

Climate activism has given me hope at a time when there is very little. We have never seen a movement like this before – a united youth movement across the world where millions of children are coordinating simultaneous actions for a common goal. This is a movement of hope, kindness and acceptance. I have never been more proud of anything in my life than I am being a part of this generation. And if these are the people I am to share my future with, then I have a lot of hope for what is yet to come.

Surround yourself by optimistic people who share your vision for a better future.

RAINA IVANOVA

Aged 15

GERMANY

The documentary *An Inconvenient Truth* by Al Gore was the moment I first realised what was really happening to the world. It pained me to see the pictures of catastrophe. That's what got me started, because just crying wouldn't help anyone.

Earlier this year, I told my little sister about climate change. She is seven years old and she loves animals very much. On hearing about the sixth mass extinction, she started to cry because she was so scared. It felt like losing the ground under my feet to see how it made her feel.

Being a young climate activist is very stressful. I get criticised very often and made fun of sometimes. The thing that bothers me the most is that now my friends feel like I'm judging them each time they drive somewhere or buy something made of plastic.

In my home town of Hamburg, the biggest issue we're facing is rising temperatures. Germany recorded the hottest summer ever in 2019. Things like this make it hard to live life as we usually do. It's harder to concentrate in school because we're not used to such high temperatures and don't have air conditioning. After school we cannot stay outside for long because it's just too hot.

Leaders should realise that an environmental collapse won't be good for their economies. If I could change one thing about my country, it would be to change people's mindset when it comes to appreciating our environment. Change starts with yourself, and if more people appreciated nature and the creatures living among us on this planet, they would protect them.

Al Gore's documentary *An Inconvenient Truth* was first released in 2006. As one of the highest grossing documentaries of all time, the film reportedly had a greater impact on public opinion about climate change than any scientific paper or report.

FEDERICA GASBARRO

Aged 25

ITALY

Three things in my life encouraged me to take up activism.

I was born in Abruzzo, a place in central Italy surrounded by nature and mountains, but I have always lived in Rome with my family. On holidays, we would often go back to Abruzzo. When hiking in the mountains, my mom would collect the bottles she saw lying on the ground along the paths and then throw them in the garbage. She used to say it is important to be educated and to respect the world around us.

In middle school, during the geography lesson our teacher told us about a missing lake that no longer exists today. It broke my heart. She said it was the fault of climate change, but at the time I didn't really know what that meant. Only after growing up and studying more, have I discovered the damage we do to ecosystems.

Two summers ago I was swimming in the Mediterranean Sea. At one point, a fin appeared in the water. It looked like a shark, but there are no sharks that come so close to the coast in the Mediterranean. But all the people retreated. In the end, it was just a poor dead dolphin. When it came ashore, plastic came out of its mouth. It shocked me and I started crying. I feel hurt even when I recall the story now.

I am lucky that I don't live in a place that has already become the victim of serious disasters. But climate change has caused flooding in Italy, particularly in Venice and Matera, in the south. In Venice, floods are common, although there has been nothing like what is being experienced in recent years, where floods have severely damaged this UNESCO heritage city. In the case of Matera, it was the first flood like it – in southern Italy this is not normal. As climate change gets worse around the world, Italy will also see many climate refugees.

We need to shift from ambition to action.

I believe that we young people are a strong force in the fight against the climate crisis and that we will prevail. We make up more than half of the global population, we will grow up with the climate crisis, and we have to deal with it for the rest of our lives. We are fighting for our survival and for our future. We are fearless warriors armed with troves of scientific research, strong social media skills and the latest technologies. We have bold ideas, endless energy and unwavering determination. So we will not stop until we win. We are the citizens of the future, we can and will change the world.

As the world witnessed recently, the youth movement against climate change has made significant progress in raising public awareness and pressuring politicians to take action to deal with the crisis. We brought millions of people to the strikes, and we have won the European Union's pledge to spend billions on climate change over the next decade. Together we were the powerful scream shouted by our Mother Earth. A scream that had been silent until now. We were able to rise and make leaders hear it.

The number of people trying to cross the Mediterranean Sea and seek asylum in Europe is rising as global temperatures increase. Climate impacts like drought and lower farm yields will probably increase and thus drive more to attempt the risky migration route in the decades to come.

WHAT WOULD YOU ARE WORRIED ABOUT

Make the climate crisis personal. Get angry about it and channel this anger into using your voice to effect change.

Laura Lock, 18, England/Hungary

In Germany we have a saying: 'Geteiltes Leid ist halbes Leid.' It means 'Shared worries are half as bad.' So share your concerns and talk about climate change.

Raina Ivanova, 15, Germany

SAY TO PEOPLE WHO
THE CLIMATE CRISIS?

The most challenging thing is dealing with the pressure and self-doubt of knowing that, whatever I do, it's not going to make a massive difference. I want to change everything and do everything, but that's just not possible. If we want to solve these issues, we need to change the system.

Holly Gillibrand, 15, Scotland

LAURA LOCK

Aged 18
ENGLAND/HUNGARY

Laura Lock wrote to her school asking for permission to join a climate strike. She requested that the school name not be shared for privacy reasons.

February 2019

Dear X,

Teenagers are not anarchic or irrational for wanting change. When, at aged sixteen, I am subject to a government that I cannot influence, I have every right to be angry and dissatisfied with its failure on climate action. The UN has declared that we have twelve years to prevent catastrophic climate change; this is unacceptable. When 100 companies make up over 70 per cent of carbon emissions, climate change is no longer an individual issue. Turning off the lights as you leave a room will not change the world. If we want to see effective results, we will require high-level political change that restricts multinational corporations, but these results are not being delivered.

Students are striking as a result of thirty years of political inaction and systematic failures. These strikes are not, as you called them, 'an excuse to miss school', but a plea to those in power to declare a state of climate emergency and include young people in the political discussion. We do not want to have to strike. All of us have forthcoming exams, massive workloads and we would much rather be in class. Nonetheless, the issue of climate change is so critical that it must supersede our educational obligations.

Politicians have failed to grasp the intergenerational nature of climate change. As most democratic terms do not surpass ten years, there is no political appetite for tackling the long-term issue with radical action. Those holding power have neglected the young and, consequently, we depend on public attention to spread our message and make our demands. We are too often criticised for not engaging in politics; therefore this peaceful civil disobedience only goes to show our dedication to the environment and commitment to our future. Voltaire's 'Il faut cultiver votre jardin' is no longer applicable; we need revolutionary advances to halt climate change.

The phrase 'societal amnesia' coined by environmentalist Lucy Siegle emphasises the myopic actions of politicians and the brutal consequences these will have on future generations.

The National Union of Head Teachers has said that it 'does not condone' pupils missing schools and that during term time students should be in class. While one can argue this entrenched line of defence, it is necessary that the strikes take place during the week. In the words of Greta Thunberg herself we 'cannot solve the crisis without treating it as a crisis'. The protests must occur on a Friday to define their nature; the global issue is now more severe than the need for an education. The universal threat of climate change must come before school. While I do not dispute the importance of becoming educated, I must argue, alongside Thunberg, that there will be no need for educated peoples to save the planet if there is no planet left to save. Caroline Lucas, a leader of the Green Party, supports this stance, stating that the protests must happen during the week to indicate that we cannot carry on as usual when faced with such a grave future.

The backlash of high-level officials has been based on short-sighted views that prioritise small national issues ahead of this detrimental crisis. Education Secretary Damian Hinds claimed that the only lasting effect would be 'creat[ing] extra work for teachers', and Theresa May deemed it a waste of time. When thirty years of politics have been wasted enhancing big business rather than restricting it, I do not think it too extreme to be demanding a habitable planet. Sustainable development should not be utopian. It is entirely achievable, and perhaps it is our youth that gives us the hope that it can be attained.

The [school name] mission statement asks us to adopt 'a sense of environmental awareness' and develop a 'responsibility to the international community'. How can [school name] endorse such principles and not support student participation in a global climate strike? Active civic engagement is our duty and striking has become necessary to facilitate change. [School name] should support its students as they fight for a better and more sustainable planet. With internal initiatives such as the Green Committee, [school name] makes every effort to appear environmentally aware, so how can you not support these global strikes?

Missing a day (or two) of lessons for the climate strikes isn't going to negatively impact any student. [School name] students that took part in the last strike described it as 'empowering', 'inspiring' and 'beyond incredible'. The first-hand experience is one that will incentivise any student to become more politically active and environmentally conscious. The experience has made me far more aware of my actions and inspired many of my peers to become vegetarian. However, as stated earlier, in this case, it is the bigger picture that counts. I urge you to see the importance of the March 15th event; surely its significance outweighs attendance figures and administrative restrictions.

By denying students the right to take part in these strikes you are hypocritically implementing a double standard on the [school name] mission statement and preventing students from carrying out their democratic rights. By encouraging students to take part you are fostering a truly international community that will manifest itself in conscientious and well-rounded students. The statistics have been available for years, the news of climate change is not new; what is new is the action. The sudden rise of the youth on the political stage only strengthens our right and demand to be heard; it is not much to be asking adults to consider our futures as well as their own.

I do not doubt that you understand the points that I am making and so I must ask, with the wealth of climate data, the myriad of public support and the incredible outcomes of participation, how can you not support the involvement of students? Moreover, if you do support it, why can't you justify the absences as 'exceptional circumstances'? In his UN address, Leonardo DiCaprio said:

'you will either be lauded by future generations or vili-
fied by them'. Climate change is a defining issue of both
my and your generation, I will not passively watch as our
planet is destroyed.

I am imploring you to make the right decision here.

Laura Z. Lock

AGIM MAZREKU

Aged 23
KOSOVO

As a teenager living in the youngest and most isolated country in Europe, Kosovo, I had the idea of organising various activities through which young people would be able to express their creativity. Together with other like-minded friends, we started organising trips to the Sharri National Park, the southernmost mountains of Kosovo. Soon we realised that these mountains were our home and that they also provide drinking water to a large region of the country.

The Sharri mountains are an essential component of my commitment towards environmental protection. I found inspiration in the mountains, but I also came to the realisation that the mountainous ecosystem is facing a threat from unjust human activities. A desire to seek justice for my beloved mountains became the major reason why I started getting interested in environmental issues and the climate emergency.

Climate change affects everyone, but it affects the most vulnerable disproportionately. Many people of my age were born during or just before the Kosovo War. External factors, political and non-political, made Kosovo a place of instability. The country hasn't fully recovered, and climate change poses risks on top of that. Many young people of my generation have had to flee the country because of high unemployment and instability in Kosovo. Droughts, temperature anomalies and floods will only increase unemployment by affecting our economic sectors, such as agriculture, industry and the service sector.

I have approached my campaigning work from many angles. My motto is that as long as I am able to have small impact on an environmental issue, I am all in. The fields that I have campaigned the most for are renewable energy and climate policy.

My main message is that the climate crisis must be considered a global emergency and it requires immediate and unambiguous action. International negotiations on a subject such as our common climate must not leave behind places like Kosovo, which are still recovering from the ravages of war.

ADRIÁN TÓTH

Aged 30
BELGIUM

In 2015, I was conducting field research on green turtles living on Pulau Redang island in Malaysia. These turtles had consumed plastic they had found in the oceans, mistaking it for jellyfish, which is one of the main sources of food for juveniles. That's when I realised that our oceans are becoming a plastic soup. Now I live in Belgium – far away from the turtles – but I want to make sure we don't pollute their homes.

The iconic Place du Luxembourg square in Brussels is situated in front of the European Parliament and is the place to be for many EU officials and bureaucrats wanting an after-work drink. Unfortunately, many of the beverages served outdoors come in single-use plastic cups. The result? Dozens of trash bags full of plastic waste with over 10,000 single-use plastic cups on Friday morning after each night of unsustainable fun.

A group of friends and I started brainstorming an initiative that would tackle our shared frustration about the situation. What started as a Facebook page encouraging people to bring their own cups, has grown into a reusable cup deposit scheme called Plastic Free Plux, founded by myself and Elias De Keyser.

Since the introduction of reusable cups in the second half of July 2018, about 100 reusable goblets were rented out every Thursday night, saving about 450 single-use cups each time. For the Plux team, this was only the start. Since then we have been working on a permanent solution together with the bar owners, reusable cup pro-ducers and the different breweries supplying the bars.

In February 2019, Café Luxembourg decided to make the leap and introduce the reusable cups inside their facilities. The initial testing phase was successful where the cups are rented out for a deposit of 1 euro. The reaction from the public was also very positive, which made three more bars switch to our system. We're now working to convince the remaining bars to switch, and ultimately make this model work across the rest of the city. So far our campaign has saved over 100,000 single-use plastic cups from going to landfills or being incinerated.

If you have the right mindset, the rest will follow.

We can all act on the climate emergency by starting to educate ourselves on the scale and urgency of the problem, and following this by changing individual behaviour in our households, neighbourhoods, communities, cities, etc. I truly believe that the necessary sustainability shift and transition will come from grass-roots movements, and by giving the spotlight to projects and solutions that are contributing to climate change mitigation and adaptation. Once these are scaled up, the systemic transformation will happen naturally.

I hope to look back in 2045 and be able to say that we've solved the climate crisis by acting swiftly and collectively, leaving no one behind.

Plastic contributes to greenhouse gas emissions at every stage of its life cycle, from how it's made to how it's disposed of. By 2050, plastic will be responsible for up to 13 per cent of the carbon budget.

AFRICA

CONTINENT: AFRICA

POPULATION: 1.2 BILLION

BIGGEST CLIMATE CHALLENGES

- **WATER SCARCITY** – Along with the Middle East, North Africa is the most water-stressed region in the world. The basin of Lake Chad, for example, which covers parts of Nigeria, Niger, Chad and Cameroon has been a water source for between 20 million and 30 million people. But it has shrunk by 90 per cent since the 1960s, due to climate change, a growing population and unplanned irrigation. With an average temperature rise of 2 °C, droughts are projected to increase and precipitation could fall by 20 per cent, further worsening the water crisis.

- **COASTAL EROSION** – Changing rainfall patterns and rising seas are accelerating coastal erosion in western and eastern Africa. Erosion rates are particularly high in Benin, with an average loss of four metres per year on 65 per cent of its coast.

- **ADAPTATION** – More than half the world's population of extremely poor people live in Africa. With few resources to adapt to rising sea levels and heatwaves, and to food and water scarcity, this population is particularly vulnerable to the effects of climate change despite contributing the least carbon emissions.

- **EXTREME WEATHER EVENTS** – Droughts and storms are becoming much more frequent, with many more landslides. The continent recorded 56 such extreme weather events in 2019, compared to 45 in 2018.

KALUKI PAUL MUTUKU

Aged 27

KENYA

Climate change has made me think about the future of food in Africa. Sub-Saharan Africa is expected to get drier and fresh water is expected to be a reason for conflicts. That is why I founded Green Treasures Farms, an initiative that works with women and the young in rural Kenya to teach them about organic farming, water harvesting and enhancing environmental sustainability through growing trees.

I have to do this because climate change does not care if I am prepared or not. I know climate change will worsen, so I have to constantly keep engaging with my community and exploring different resilience mechanisms.

It is without doubt a fact that Africa, the continent that has contributed the least to climate crisis, suffers the worst impact. Kenya is no exception. We have experienced extreme droughts, famine and even delayed rains.

Forest fires have occurred in major forests which contain most of the country's biodiversity. Just recently, floods have hit Nairobi, the city I live in, hard. People have succumbed to the flooding, buildings have collapsed and livelihoods have been lost.

I have seen just how environments can change from serene states to ugly and barren landscapes. I have seen how once roaring rivers and magnificent waterfalls can disappear in the space of a few years. Now I want to make the future better and more beautiful.

I campaign for the meaningful inclusion of young people and indigenous and front-line marginalised communities in the processes of designing and implementing climate solutions. Nature-based solutions deal directly with the very environments these groups have first-hand knowledge of, and yet these people are so often disregarded or treated with disdain by political leaders. With 350.org Kenya, we are also pushing for 100 per cent renewable energy transition in Kenya.

It is okay to be worried. We should all be worried. But worrying is not enough. We have to act, and act fast. The clock is ticking and we are running out of time. Countries need to make every effort to combat climate impacts – this is a war we can win. They need to see young people as a huge asset, to innovate and find creative solutions for our common future.

My little thing is giving nature a voice. I speak up for nature, and I speak up for biodiversity, for our own life support system. We all must do our little things, because those little things will lead to great results.

My mother, a single parent, has always been supportive of my dreams and my engagement. She also believes in climate science, and wants a change that is good for people, for nature and for the planet. She often tells me how proud she is to see me leading and fighting for the environment.

Young people know that world leaders have been lying to us.

Some leaders perpetuate lasting stereotypes about activism: that it is for idle, unemployed and chaotic people. It is worse if you are a youth activist from Africa. No one takes us seriously. I have studied natural resource management in university, but even that is not enough.

The media is to blame, too. We have often been portrayed as chaotic people, which perpetuates the stereotype. Western media, especially, does not consider 'African' youth activists as punchy enough to share their stories with the world. Youth activists from the developing world deserve to be heard, and their efforts deserve to be recognised.

Politics is a subject we cannot evade. Whether we like it or not, everything we do in our society revolves around the politics of the day. Without political goodwill, we cannot do very much. So, as I push for climate action in Kenya, I wish I could change the political class to include those who are accountable to the people and are leading the fight to address climate change.

NCHE TALA AGHANWI

Aged 25

CAMEROON/USA

Our planet is changing before our very eyes. It would have been so comforting if these changes were making the planet more comfortable for the habitation of bio-diversity. Sadly, this isn't the case.

Climate change is without doubt a major problem in Cameroon. Farmers have been hard hit. The climate crisis has forced grazers to stray far from their usual grazing grounds into areas where crop farmers are struggling to adapt to the unpredictable weather pattern. The result is recurrent farmer–grazer and inter-tribal conflicts. These climate troubles are exacerbated by the civil war between the anglophone and francophone regions. When young men and women are subjected to horrific violence, farms are destroyed and whole villages are burnt to the ground, it is hard to see climate change becoming a priority. Unless there is a return of peace in Cameroon, we will fail to deploy climate solutions.

If I could change one thing in Cameroon, it will be the present regime, which has failed for thirty-seven years to implement concrete policies to address poverty, climate change and the respect of human rights.

* * *

I was having a regular WhatsApp chat with a good friend in Ethiopia. After talking about everyday issues, she began to show interest in my work as a global climate activist. She said, 'I want you to be honest with me. Are you doing all these things because you are really worried about our future and our planet or are you doing it just to be on track for a good career?'

For a split second I paused in shock. It wasn't because I haven't had these questions posed to me before. I was shocked because people keep seeing the work of climate activists as an economic venture or simply as a career path. This is evidence enough that most people are blind to the realities of a changing climate and its devastating repercussions.

There is so much at stake. Our immediate environment as we know it is at risk of becoming unrecognisable, and our biodiversity risks being lost. Several iconic species have disappeared and it is already clear to scientists that more will disappear. The comprehensive United Nations report on the state of biodiversity paints a stark picture.

Our ignorance and refusal to embrace climate science has further exacerbated the problem. The science of climate change has been clear for many years now: our planet is warming and we are to blame. It has quantified the degree of urgency. It has given us a time frame for irreversible changes.

All this is scary to digest, but science has never limited itself to painting a doomsday picture of our actions. Science has given us hope. It has provided ways in which the climate crisis can be arrested.

That is why I became an activist and founded the Africa Science Diplomacy and Policy Network. In less than a year, we have put together a strong, committed and diverse team of 700 young climate activists across the continent of Africa. I have led my team members across the world, adopting a bottom-up approach to tackling the global climate emergency. I have also insisted on the necessity of linking policies to evidence. I have organised several workshops on climate change and environmental sustainability, led climate walks and taken part in climate strikes.

Working in close collaboration, my team has adopted nature-based solutions as one of the effective ways of tackling the climate and environmental crisis. We have planted more than a thousand trees since I launched my organisation in July 2018. Using local radio programmes, we have been able to reach out to thousands of people.

The clock is ticking, but you can make a difference.

SEBENELE RODNEY CARVAL

Aged 30
ESWATINI

I am from a country that is highly vulnerable to the effects of climate change and which needs to act urgently in building resiliency.

We are already experiencing higher temperatures than normal in my country, and it has affected my health. High temperatures have caused me dizziness and headaches. I now have to use air conditioning, which I never used in the past. Schoolchildren have also been affected by heatwaves, which have caused some of them to faint.

We have also experienced water shortages, and we can go days without receiving water from the utility company. Storms have occurred with greater frequency too, and they have affected my house, vehicle and garden.

Poverty in Eswatini is high, and climate change has made the situation worse. Most Emaswati, as the country's people are called, reside in rural areas where they practise subsistence farming, which is highly dependent on rainfall. The animals they raise have died because of frequent droughts. The country has been losing a lot of money because it has to provide food subsidies to affected citizens.

The vulnerable innocent citizens of Eswatini push me to work hard. I am a project coordinator working on climate finance, assisting them to adapt to the adverse effects of climate change. I'd like to see more investment in low carbon and climate-resilient projects in this country, and the creation of an environment for such investments to attract those in the private sector.

But having your voice heard by decision-makers is challenging when you are young. Young people do not get offered a platform to address country leaders. We need assistance from the developed world in order to effectively adapt to the consequences of climate change. Assistance in the form of technology, knowledge and, of course, finance.

We all have a role to play. Choose your part.

Who must pay the price for climate change? Is it vulnerable people with the limited ability to adapt, or should it be the industrialised world that has been emitting and causing this change in the climate while improving their country's economies?

Climate change is happening and we need to act fast. If we work together we can achieve our climate action goals.

JEREMY RAGUAIN

Aged 26
SEYCHELLES

I am doing my utmost to highlight three major points about climate action.

Firstly, I think a lot needs to be done to emphasise and explain the inequity of the climate crisis, in which countries like Seychelles are the least responsible yet most vulnerable. I try my best to support Seychelles' position, which aligns with that of the Alliance of Small Island Developing States in terms of raising ambitions and highlighting the need for climate adaptation finance. The climate crisis violates the human rights of people like myself, young African islanders from coastal communities, who will become stateless if business continues as usual.

Secondly, I am talking directly to politicians and policymakers in Seychelles to stop the exploration and possible exploitation of oil or natural gas in Seychelles' Exclusive Economic Zone, taking a clear stance in advocating for a moratorium on such activities, like Costa Rica and Ireland have committed to. Drilling for oil or gas is like drilling a hole in your boat when you are in the middle of the sea.

Thirdly, I have been pushing for the word 'Anthropocene' – the idea of an epoch shaped by humanity – to be better understood and used by young and old. There is a chance that plastic pollution runs the risk of distracting us from the urgency of the climate crisis. But I see the momentum behind the need to tackle plastic pollution issue as ultimately positive. It leads to us seeking recourse to conservation and environmentalism. Using the concept of the Anthropocene, I can highlight plastic pollution – as the most visual, easily understood symptom of our consumer society – in order to help people understand more widely humanity's impact on the planet.

Finally, I also work with grass-roots and youth organisations to educate and share resources to allow people to have opinions and take action.

* * *

I grew up on a farm in Seychelles, and that has meant I have always felt connected to plants and animals. I love being outside, being able to snorkel and fish.

Right now Seychelles is experiencing extreme weather events that affect my work. I co-coordinate the Aldabra Clean-Up Project, which removed more than twenty-five metric tons of marine plastic litter in a five-week-long expedition. Friends and colleagues working on Aldabra are exposed to greater and more frequent storms, which delay supply flights and endanger people's lives. Meanwhile the main islands on which most Seychellois like myself live are seeing high tides – made worse by sea level rise – which block roads, while extreme weather can cause damage to homes and critical infrastructure.

Seychelles may not be habitable in the next couple of decades. Not only because of sea level rise and extreme weather events increasing in frequency and intensity, but because ocean acidification and increased ocean temperature will kill the coral and fish that are vital to our economy – coral bleaching is already taking place. Both artisanal and industrial fisheries will be drastically affected, and the latter is 80 per cent of our exports – if tomorrow we stopped exporting tuna because they are not being caught in our waters and landed at our port, the price of living would double. This would make living in Seychelles impossible for most, including me.

I think about having kids and I come to the conclusion that I cannot. I don't want them growing up in a world that may be devoid of life in the ocean, where you can't play outside because it's dangerous and food and other things we have taken for granted are no longer as readily available. I think I could become stateless and unable to have kids or pass on my identity, culture and way of life. If left unchecked, climate change will eventually leave Seychelles a shell of what it used to be.

Don't just let the news wash over you, do something to save yourself.

LESEIN MATHENGE MUTUNKEI

Aged 16
KENYA

My activism started when I was thirteen years old and I read an article about how people are cutting down trees and destroying the forests at a shocking rate. It said Kenya is losing forests the size of six football fields every day. It explained that the loss of trees worsens climate change, by disrupting the water cycle and bringing more drought and floods.

I was shocked that this serious effect of climate change is not known by many young people and yet we are the ones who will suffer the most. This made me feel very sad and I wondered why people destroy our forests and our environment when they know how this will harm us all.

That's when I decided I must do something to protect our environment, no matter how small it is. I also decided I needed to learn more about climate change since we are not taught it in school. As an avid footballer, I came up with a creative and motivational way to score more goals and help the environment at the same time. I called it Trees For Goals and committed to planting a tree for every goal I scored.

I introduced Trees For Goals to the team at my football club, and they joined me in ensuring we score more goals and thus plant more trees. During training I share information on climate change with my team and the need for us young people to take action. I noticed that many of them had never planted a tree, and so I decided I wanted to get everyone involved.

When I shared the plan to plant 600 trees in one day, my friends laughed and didn't think it was possible. When the day came, I was a little nervous and thought maybe they wouldn't turn up. They all turned up, some with their parents, siblings and other friends – despite the morning drizzle, everyone was ready to get to work. A conservationist explained where we needed to plant indigenous trees and demonstrated how to properly plant a tree. By the end, we had planted all 600 in an hour, and it made me so happy to see my friends naming their trees and being so engaged with the environment.

Now, I spend time visiting other schools and sports clubs to talk about the environment and encouraging them to start their own eco-initiatives.

One of the most challenging things about being a youth activist is where to get authentic, accurate, up-to-date and easy to understand information on climate change and the actions we can take as citizens and young people to reverse the negative effects. As a student, juggling school work, football training, friends and running my Trees For Goals project can also be challenging.

* * *

My country has been severely affected by the disruption to the rainy season. When visiting my grandparents in the rural area, one of my favourite places used to be by the river which passes by their home. Now when we visit, we find the river is often dried up. My grandmother told me this has affected the crops in the region. Farmers who depend on rain to grow crops are struggling and many Kenyans are starving.

We now get water in our home on fewer days in the week than before. My mum told us that the level of water has fallen and so the city has started rationing water. Sometimes we go three or four days a week without a supply of water. In the future, climate change will reduce the amount of water, making our taps dry, and that will make it even more expensive to buy water. This will force the government to ration the water people can use, like what happened in Cape Town in South Africa. This can create fights among neighbours, because we cannot live without water and people will fight for their lives.

When it does rain, the heavy rain floods the river near my school, making it impossible to enter. As well, it carries toxic waste from the riverbed. These floods damage crops and cause mudslides, which have destroyed villages and killed many children and people. Our generation will be most affected if climate action is not taken, so giving up is not an option.

Since I started Trees For Goals, I have learnt so much about climate change and the current global crisis. I have also seen so many of my friends and other young people ready to listen to what is happening and what they can do. Best of all, I have seen them realise they do not have to be a grown-up to take action.

Twende kazi – let's get to work!

My greatest inspiration has been the late Nobel Peace Prize winner Professor Wangari Maathai. This woman was a trailblazer, an iron lady and a true daughter of the soil. I always call her my environmental mother.

Kaluki Paul Mutuku, 27, Kenya

My mother, Elizabeth Sadler. She taught me the value of standing up for the things I believe in and being me. As a single mother of two, an artist and a survivor of an abusive marriage, she proved to me what is possible with hard work and determination.

Jeremy Raguain, 26, Seychelles

INSPIRATIONS?

My parents, Samantha Pearce and Mark Sampson, my climate activist friends I have grown to love and admire, and the internationally known activists working tirelessly for change. These people are the light at the end of the tunnel when I need it; they are the people I turn to for hope; they have inspired me to continue to fight for climate justice.

Ruby Sampson, 14, South Africa

TOIWIYA HASSANE

Aged 21

COMOROS

I come from a small island state that is also a biodiversity hotspot. With no industrial activity, the main factors impacting on environmental degradation in this country are poor waste management, deforestation and water pollution. I am campaigning for the protection of the forest and for better waste management.

I studied life sciences, which enabled me to learn about the importance of biodiversity in Comoros. This training also made me realise the importance of conserving species to keep the entire ecosystem stable.

In Comoros, the fight against climate change is a difficult task because people have not yet been made aware of the risks associated with the climate. Unfortunately, there are still Comorian citizens who are convinced that the forest is much too thick and that it is necessary to reduce it. It is obvious that our most urgent problem is ignorance of the danger. The first step to be taken is therefore to raise public awareness of climate issues.

But Comoros is already affected by climate change. The seasons are out of control, species are disappearing and sea levels are beginning to rise. The problem faced by small island countries is likely to get worse.

If I could change one thing in my country, it would be the priorities of our leaders. We have the means to fight the crisis, but we are losing the fight. It is as if our house has caught fire, but we are busy filling cans with water for provisions.

* * *

The biggest difficulty I have faced as a climate activist is cynicism. I am often confronted with questions like 'What do you know about the future?' 'Do you think you are God?' 'How do we know it's not just one of your policies to get money from us?'

I think this is because climate change has been politicised over the years. It is time to find new ways to stop this crisis that is getting worse every day. It is time to find a way to counteract all the damage caused by industrialisation, while ensuring that we do so at a pace that will not cause popular uprisings. We must innovate, and I believe it is young people who are able to provide the innovative solutions.

It is young people who have enough ambition to persevere in the search for new alternatives, enough strength to withstand the risks that this will involve, and enough naivety to believe in it. Because sometimes all we need is to believe that the world belongs to us, to be able to consider all the possibilities, otherwise we have far too many restrictions that prevent us from emerging with sustainable results.

Comoros is situated some 300 kilometres off the south-east coast of Africa and is made up of three islands. Rising sea levels pose a huge threat to the islands. The entire population lives within ten kilometres of the coastline.

KOKU KLUTSE

Aged 28
TOGO

After studying agro-economics, I started my own farm in a village located 85 kilometres from Lomé, the capital of Togo. I love the land and the environment so much. I have an estate of 2.5 hectares and produce fresh tomatoes, green peppers and eggplant. Everything was fine until October 2014.

The off-season crops were at flowering stage when, in November 2014, heavy rains hit the region for a whole week, flooding the entire field and making my investment worthless. Now I was heavily in debt, but the fate of neighbouring producers who suffered the same blow was more worrying. Since then, I have made it my mission to do something to ensure that farmers are free from flooding in the future.

One of my main ideas was to solve this flooding problem. Southern Togo has cut down all its forests to produce charcoal, a product found in more than 95 per cent of homes in Lomé and other cities in the country. Charcoal is the main fuel used in Togolese kitchens. The destruction of flora and fauna was one of the main causes of rainfall disruption throughout the West African sub-region. Something had to be done about it.

In my analysis of alternative sources to charcoal, I realised that butane gas was quite accessible, and that if promoted, little by little, the situation of 'zero charcoal' could become a reality. As a result, parts of our forests could be conserved instead of being sacrificed to the charcoal producers. Togo was one of the few countries in Africa that did not have a butane gas station; not even a single station where people could go to get gas.

Although burning of fossil fuels contributes to climate change, in the short term, until poorer regions can afford electric cooking stoves powered by reliable renewable electricity, it's better to use natural gas, petroleum gas or butane as a fuel source instead, to help prevent deforestation and cut indoor air pollution, which kills more than 4 million people each year.

I sold my bike, which was my sole mode of transport, to found the Jony Group. In less than five years we have provided more than 1.5 million tons of butane gas to about 200,000 households in and around Lomé. The company currently employs forty-two people directly and seventy indirectly. Through its promotions, the Jony Group has reached tens of millions of people with the message of environmental protection.

TSIRY NANTENAINA RANDRIANAVELO

Aged 28
MADAGASCAR

The island of Madagascar is amongst the ten most climate vulnerable countries in the world. During the last two decades the effect of climate change has become apparent in the form of extreme heat, floods, droughts and violent cyclones.

In the last few decades Madagascar has lost 90 per cent of its forest and 1 million hectares of arable land. The capital of the country is overpopulated because of climate immigration, insecurity and malnutrition. Every day I see people queuing for a jerrycan of water that they buy with a quarter of their daily income. In the future, without concrete climate action, I worry that the sea level will continue to rise and resources like water and forests will continue to disappear.

Even though young people make up the vast majority of the country's population, we are not included in any climate action. I frequently face ageism because I'm young. Leaders need to involve young people because we have a voice. I urge everyone to take at least a small action against climate change. When our efforts are joined, our small contributions will make a big impact.

Madagascar is a biodiversity hotspot, with 90 per cent of its wildlife found nowhere else on the planet. But climate change threatens 25 per cent of its species.

RUBY SAMPSON

Aged 19

SOUTH AFRICA

When I was eleven years old my parents took me around Africa clockwise. I saw first-hand the dramatic effects of climate change: floods in Sierra Leone, droughts in South Africa and desertification in Senegal. I saw people dying, and I came back to a society blinded by profit, too greedy to accept the harsh reality of climate change.

Before I went on this trip, I had plans and goals, a future not ruled by the climate crisis. My dreams were crushed by the realisation that we as a species are slowly but steadily killing ourselves and the creatures around us. Since then, I have felt this crushing obligation and responsibility to act, because no one else is.

I have memories of living in severe drought in the deserts of Namibia where there was nothing left to feed the cattle, and so the herdsmen fed them cardboard soaked in brine in order to fool their stomachs into feeling full. Memories of grief for those who died in the mudslides in Freetown, Sierra Leone. Memories of sympathy for close friends who developed permanent skin discolouration from polluted water, or friends whose families have no money because the crops have failed, again. All of these memories have left their mark on me, pushed me to work harder, prioritise climate action and strive for climate justice for those who deserve it.

The African Climate Alliance, the non-governmental organisation I co-founded, has four main demands that it has put to the South African government:

1. Declare publicly that we are in a climate crisis.
2. Put a moratorium in place on all new coal, gas and oil mining licences.
3. Convert the electricity sector to run on 100 per cent renewable energy by 2030.
4. Create a mandatory education curriculum about climate change and its effects on South Africa for all schools.

If I could change one thing about South Africa, it would be the lack of education. Apartheid forced generations to go uneducated, unemployed and penniless. That system mercilessly categorised people on the basis of race, giving opportunities to a few while the rest suffered with nothing. Decades of such abuse doesn't disappear with the coming of democracy.

I am protesting for the people suffering now, not only those who will suffer in the future. People are feeling the effects of the climate crisis already, in Mozambique, Sierra Leone, Senegal, Namibia and South Africa. The families without homes, the children without parents – they are who I am fighting for. I am protesting for recognition of the emergency we find ourselves in. I am fighting for climate action – for education about the climate crisis and for climate justice and system change, for resources and funding to be given to struggling farmers and their families – a just transition away from non-renewable energy and most importantly no more gas, coal or oil.

Mitigating catastrophic climate impacts and tackling joblessness, inequality and poverty are the same fight. It's time to stop telling South African workers they must cling to insecure, unsafe and unhealthy jobs in the coal industry, and to offer them sustainable employment as part of a just transition away from fossil fuels and towards renewables. The government's inaction in the face of the climate emergency is jeopardising the lives of South Africans.

Don't let your fear stop you, let it unite and drive you.

TAFADZWA CHANDO

Aged 23

ZIMBABWE

I grew up in a society with local leaders who did not care for the environment. Water was scarce. I wanted to find solutions to improve and empower our communities.

When I was working on a conservation project in 2019 in remote areas of Zimbabwe, I was shocked that most people did not even know about climate change. It prompted me to petition the local government to raise awareness, and I created clubs in local communities and schools to raise awareness.

My activism began when I was in school, where I headed our environmental club. We built greenhouses to grow plants and brought in professionals to talk about recycling.

My parents saw my interest and they have been supportive. They even provided early funding for the youth club I founded.

* * *

It's hard to get attention when you are young, unless you take it on to the street. But the political situation makes it difficult to do that. The Zimbabwean government sees protesting or activism as a threat.

In 2019, the government passed a bill that makes it hard to get permission to protest. We organised climate marches but none were approved. That is why we are taking our fight to the courts. We have seven pending cases on wetlands protection.

The only rent we can pay to stay on this planet is through activism.

DELPHIN KAZE

Aged 25

BURUNDI

I live in a country where agriculture is the main economic activity. Climate change is already affecting harvests because of changes in the rainy season and increased occurrences of droughts and floods.

This is a developing country, and the government doesn't have enough resources to help the agricultural sector adapt and become resilient. Economically we are suffering, as we cannot help those who are affected by the damage that floods cause or feed those who have nothing to eat because of the droughts. Climate change is really a social matter. These struggles are proving to be obstacles to achieving sustainable development, but if nothing is done, lives will be lost to famines and floods.

* * *

My campaign is against the alarming rate of deforestation, which is caused by wood being used as the main source of energy for cooking in Burundi. I founded a social enterprise which promotes the use of clean cooking through the use of eco-friendly charcoal, which is produced from organic waste such as maize. The eco-friendly charcoal doesn't produce smoke and thus it also reduces indoor air pollution. Its use helps protect people's health, especially the health of women and children.

Being a youth climate activist in Burundi is not easy. Our ideas are not given the same consideration as those of our elders.

Nine out of ten Burundians are employed in the agricultural sector, so climate change will affect a huge proportion of Burundi's population.

ELIZABETH WANJIRU WATHUTI

Aged 24

KENYA

People protect what they love. I still remember spending my childhood in nature at its most pristine. All I ever knew and saw on my way to school were trees ahead of me, bushes beside me, whistling winds around tree trunks, clean streams that were flowing close to my homestead, and just that special feeling of peace and tranquillity in harmony with nature. Now I dream of a world where humanity will stop threatening our life support system and of the only place I have always called home, that is, nature.

I am an environment and climate activist and the founder of Green Generation Initiative. Through my initiative, for the past two years, I have been nurturing children and young people to love nature and be good environmental stewards by involving them in practical environmental education programmes and a tree-growing campaign that enables them to learn by doing. I run a campaign dubbed 'adopt a tree' which aims to ensure that every child in every school gets an opportunity to plant and adopt a tree. The initiative also aims at addressing food insecurity through the establishment of food forests in schools, where we plant mixed types of fruit trees in a designated corner within the school compound. The initiative is mostly focused on greening schools, introducing practical environmental lessons, and inculcating in people a tree-growing culture to help increase forest cover while tackling climate change.

Kenya is very vulnerable to climate change, with current projections suggesting that its temperature will rise by 2.5 °C between 2000 and 2050. I have witnessed unusually heavy rainfall causing floods, mudslides and landslides leading to the deaths and displacement of people. It is terrifying to see people being carried away by floods, lives being lost, children suffering from more and more respiratory diseases and even going for days without food due to increased food insecurity. Children and women are the most vulnerable.

The largest carbon emitting countries are still blocking potential opportunities for African nations to access finance for adaptation and mitigation. This global betrayal and selfishness must come to an end if we are to avert future catastrophic events caused by the climate crisis.

NDÉYE MARIE AIDA NDIEGUENE

Aged 24
SENEGAL

They were playing along the beach with a ball at their feet and fury in their eyes. They were playing on this strip of land, between heaven and earth and between earth and ocean. The Atlantic faced them, immense and impetuous. They played on this strip of land, which year after year shrank under the helpless gaze of the young boys and girls who frequented it every day.

Today, it has disappeared in the total indifference of a world which, no doubt, had forgotten that each piece of soil, just as all the pieces of planet Earth, are equal.

I started this text without speaking of context. You are in Kayar, a small town located 58 kilometres from Dakar, the capital of Senegal. Kayar is on Senegal's coast, which stretches over 700 kilometres. The ocean has always been the main source of income for this fishing community. Kayar is famous for its fish market and its coast.

Let's go back to the young people on the beach, to the young girls splashing their feet in the water, to the rustling of the waves that fall at the foot of the continent. The ocean has turned against its sons. Around the world, islanders and coastal populations tremble with dread in the face of rising sea levels. There is not a day that goes by without talk about crumbling coasts, disappearing islands, floods, climate refugees.

The Kayarois were once the masters of the sea. They tamed it from generation to generation, but today it is out of control. The boys' soccer field was cut in half in less than ten years. The pirogues, traditional boats used by Senegalese fishermen, are no longer kept on the coast. There are no more photos of canoes lined up as far as the eye can see facing the ocean. Today it is necessary to bring them in further, and alas it is not uncommon now to see canoes pushed by men onto the land and sheltered.

And if you take the time to stop, if you take the time to ask them, they will tell you, 'The sea is rising, our canoes, and even our homes are in danger.' Kayar is in danger. Faced with the climate emergency, faced with the rising waters, constant flooding, coastal erosion, the population of Kayar remains powerless. Nothing protects people from the fury of the ocean.

In Bargny, 31 kilometres from Dakar, the pattern is the same. The sea advances silently. Local populations, powerless, see their homes disappear. Their houses collapse in the face of total indifference. There is no policy in place. We prefer to close our eyes, to focus on the other world. The world of commitments, major reforms, major projects capable of saving cities.

Talking about the disastrous effects of climate change without talking about Saint-Louis of Senegal and its emblematic Guet Ndar would be a sacrilege.

Guet Ndar is a fishing village located in Saint-Louis that was flooded, creating real climate refugees. The sea, friend and main resource of this fishing village, once again turned against its sons. Guet Ndar was overwhelmed. The cemeteries were invaded by the waters, the houses collapsed, and nothing was done to help.

Is there a two-speed climate reality? Are there two planets? Or are we the pieces of one and the same planet?

Why, then, is the situation on the Senegalese coast not worrying? Why are our climate refugees left stranded, and why is their situation totally ignored? Have you thought about the future of my country? We, the southern countries, produce less than 1 per cent of global gas emissions but we are suffering the full brunt of the disastrous effects of climate change.

So are we collateral damage? Do they not realise that we are all part of the same world?

ANTARCTICA

CONTINENT: ANTARCTICA

POPULATION: 0 PERMANENT RESIDENTS

BIGGEST CLIMATE CHALLENGES

- **ACCELERATED TEMPERATURE RISE** – Parts of Antarctica are warming three times as fast as other parts of our planet. Antarctic glaciers are losing ice faster than they are gaining it, contributing to rising sea levels.

- **HABITAT LOSS** – As temperatures rise, so ice-dependent species could be under threat of extinction as their habitats are lost – emperor penguins, for example, need surface ice to breed.

- **PROTECTING KRILL** – Krill feed on phytoplankton and find food and shelter under the sea ice. Practically all the wildlife in Antarctica depends on krill as a main food source, so disappearing ice sheets pose a threat to the entire food chain. Krill are also a key contributor to the Southern Ocean's role as one of the world's largest carbon sinks.

- **OVER-FISHING** – Fishing in Antarctic waters is regulated under an international treaty. But that hasn't fully stopped illegal, unregulated and unreported fishing activities, which threaten the biodiversity of the continent.

ZOE BUCKLEY LENNOX

Aged 26

ANTARCTICA/AUSTRALIA

Zoe Buckley Lennox travelled to Antarctica on the Greenpeace ship, the Arctic Sunrise, *to protest against krill fishing in Marine Protected Areas. The following are excerpts from a daily account she kept during her time there.*

Thursday 08–03–18

Today was a good day. We finally set off towards Antarctica. I saw an elephant seal, fixed my boots and helped with some ship bits. It's also International Women's Day – this seems like an extra pertinent day to head to Antarctica, given that there is a rough history of sexism on the continent. Originally women weren't allowed on the land mass because it was believed we couldn't handle 'the extreme temperatures or crisis situations'. It wasn't till the '60s that women were allowed to work in Antarctica, almost a hundred years after the first explorers arrived. Apparently it's because there were no hairdressers or shops so we wouldn't be able to survive, or have anything to do. Because, you know, science is of no interest to women. *Gah.*

* * *

Saturday & Sunday 10/11–03–18

It's getting colder outside, the temperature drops every few hours as we get closer. The Drake passage only served us a two-metre swell and the albatross have started to appear. Royal and wandering albatross are the biggest (3-metre wingspan!), and if you brave the cold and wind outside you can probably see at least two. They soar behind the ship, effortlessly swinging on currents of wind, touching the tips of their wings on the water and moving around the waves. They are kites with invisible strings attached to the end of the ship, catching the updraught as we chug along. Skuas and petrels also join in. Petrels range in size and shape. Some with 2-metre wingspans, others with around 23-centimetre wingspans. These are storm petrels, and they dance on the water and flutter about. They are beautiful with their bow-shaped wings, legs perpetually prancing, black coats and white chests. I saw penguins today too! Just for a moment they leaped out of the water and slid back in, four or five in a row.

* * *

Wednesday 14 – 03 – 18

Today we moved to Paradise Bay and did a drive-by to check out some of the fishing vessels down here. They are mostly krill vessels, with huge nets that scoop up millions of little pink krill. In some of the photos I've seen the water welling up behind the ship is pink from the high density of krill. They are k(r)illing in the name of growth, in the name of omega-3 supplements and fish meal. Krill are hugely important and protecting their habitat means protecting a whole ecosystem. Reducing krill fishing can also help to buffer the effects of climate change – krill help bind up a heap of carbon that would otherwise flow back into the atmosphere. The international body that controls the waters around Antarctica is CCAMLR (the Commission for the Conservation of Antarctic Marine Living Resources) and at the moment they allow krill fishing, even though it's officially a Marine Protected Area. But the rate these pink friends are being scooped up is enormous and given habitat loss (via climate change) and increases in whale populations (*yay*) it's just too much to be taking them out of the ecosystem.

The krill vessels are huge and they bring these things called reefers down too, in order to transfer the krill catch of one boat into a bigger ship to maximise catches. There was a fishing reefer linked up when we saw them – they had probably just collected a load – in this beautiful bay, surrounded by huge glaciers, amazing rocky cliffs and colonies of Adélie penguins. One of our boat drivers noticed that they were dumping waste into the ocean next to them, and a bunch of birds hung about picking up the organics as they floated in the water. Another of the crew picked up a plastic package floating in the bay.

On our way out around the bay we had to keep watch for penguins, seals, icebergs and whales, shouting and pointing wherever we saw them to ensure the boat driver avoided them – that's how many there are. The water is truly brimming with life.

* * *

Thursday 22−03−18

We finally got to do what I came here for.

I woke up foggy eyed and headed to the mess, and when I got there tension was high. The boat crew were being briefed on how to stop a Ukrainian fishing vessel from linking up to a massive reefer, by using our RHIBs (small high-speed inflatable boats) to get in the way.

Within half an hour they were on the water, taking photos and getting in the way of the transfer.

After seeing that the reefer and fishing vessel were linked up, and checking all my climbing equipment, we got word that we should try to do the banner first. As soon as all the equipment was ready, we loaded into a boat and zoomed straight for the vessel. As we got closer I could see there were almost twenty men on deck. We used a hook and a pole to clip a climb line up. It took three attempts to get the first one, and the men on deck had knives and they cut the line before anyone could get on it. Then finally we got a good anchor and Sarah jumped on the line. Soon, with another hook, we got Meena up. We had planned the distance to be ten metres between them but in the end it was more like twenty, so a few extra boat ropes were used to extend the pull line. We managed to get a line between the two and they hauled the banner across. It looked beautiful. Huge ships, colourful banner, two little climbers and the massive cliffs of ice and mountains behind. It was almost too epic to take in. We stayed for about twenty minutes, while the on-board media got pictures and a drone was flown over. After the time was up we grabbed the banner, both climbers came down quickly and back into the boats, then with the pole we clicked both hooks off and grabbed them. A full retrieval.

We had a 45-minute turn-around back at the ship. I quickly chewed down some food and then me and Ronni got all the equipment ready for the pod. The plan was to get a rope up the stern of the ship, just above where they haul in krill. I would climb up first, drop another line, haul up a chain block (a massive chain to pull up heavy objects), the pod would be put in place below, then Ronni standing on top of the pod would hook it up and we would hoist it as high as we could. As we approached we couldn't see anyone on deck, but after we clicked the first line on and I started climbing, I was in the middle of saying 'I'm all good' when, according to onlookers, a hooded man very quickly slashed my line. I was in the water only for a few seconds before being pulled out back onto the RHIB. My dry suit held tight and I was okay, just a little shocked. We pulled the rest of the line back in, and tried again, but this time somewhere a little less reachable.

Once I was back in the RHIB we sped around to where the anchor sticks out at the bow under the deck. Ronni was up on the chain and rigging another climbing anchor, and I was going to jump on top of the pod, click the end of the chain block on and pull it up. There was another way to make this occupation happen. When I got onto the pod, which was held alongside the RHIB, it was moving like a bucking bull; up and down with the swell. Eventually I managed to get the shackle attached on top of the pod, on the hook Ronni had lowered down. There is not much room on the pod and trying to hold on when it moved so violently wasn't easy.

I started to haul the pod up slowly, one hand and then the other, grabbing this slippery, wet, cold chain and heaving. My hands had been numb and as the feeling came back they stung. Eventually I pulled enough chain through to hoist it up off the water. We stood there for a second, before high-fiving and catching our breath.

The pod is equipped for someone to live in for almost a week. We climbed back out and were on top for about twenty minutes before being told that they were threatening to go to sea, where the swell is way too big to stay in the pod. It would have reached the bottom of the pod, and it would not be safe. We thought they might be bluffing.

It took a minute to realise we were moving. I only noticed because I could see the RHIB starting to have to work to keep alongside us. By this point we had to figure out how to get out of there as quick as possible. The only option was to jump and be picked up.

We had been trained about jumping in the water and I had already gone in the drink once today so I knew I would be fine. Still it's not the most comforting situation. I jumped, plunging into the ice water, resurfaced and the boat crew grabbed me, heaved me up and we pulled away.

1, 2, 3. Ronni cut the pod. It hit the water with a loud *whoooshhh* and then bobbed for a while before it started to float away. Within seconds Ronni splashed into the water. We grabbed him and pulled away from the ship. We were safe. The pod floated about fifty metres away and the ship headed off. The RHIB quickly dropped us back at the ship to get warm, and went back to retrieve the pod. We got out of our gear, hugged everyone on the ship and cracked a beer. Surely they had got the message.

OCEANIA

CONTINENT: OCEANIA

POPULATION: 42 MILLION

BIGGEST CLIMATE CHALLENGES

- **DRIER, HOTTER TEMPERATURES** – In 2019, Australia experienced the most severe bush fire season on record, with an area roughly twice the size of Belgium burned. Lower rainfall is one reason why Australian bush fires are getting worse.

- **RISING SEA LEVELS** – Sea levels are rising by a global average of three millimetres a year, but in the western Pacific sea levels have been rising by up to twelve millimetres a year since the 1990s. Coastal erosion as well as saltwater intrusion from the flooding devastates farmland and is already rendering parts of countries such as Fiji uninhabitable.

- **FLOODING** – About two-thirds of New Zealand's population lives in areas prone to flooding. Extreme heavy rainfall is expected to occur more often, by a factor of up to four. Flooding is already New Zealand's most frequent and, after earthquakes, most costly disaster.

LOURDES FAITH AUHURA PAREHUIA

Aged 18

NEW ZEALAND

I may not live on the Pacific Islands but I feel such a strong connection to the people living there and the things they go through. Knowing the risk to my family's home guides my decisions about my future and where I put my effort.

Instead of going to university next year as I had planned to, I am taking a gap year to work with more grass-roots groups that I know can make a real difference in this climate crisis. I know this is the right decision and I am proud of what I will be doing.

My campaign involves trying to motivate my school to become more sustainable, because I believe everyone who has the power to make change happen should do something. It's especially important when there are small island nations that do their very best, but still suffer the consequences of climate change.

One of my most valuable memories was seeing 4TK, a grass-roots group that started in South Auckland, show up at the Auckland city climate strike. I heard a hearty *cheeeeeeehoooooo* – a Polynesian greeting – from down the street and sprinted to the other side of our makeshift stage to see the flags of all our Pacific Islands.

This was not a typical sight for central Auckland, because most Polynesians have moved out of the city. Seeing my culture and life brought into the heart of our city, into the space where I had been putting all my effort, filled me with so much love and pride. Finally, this space looked the way it should.

There are always more people who care than we think.

In Aotearoa (the Maori name for New Zealand) we are lucky enough not to be facing harsh fires or suffering from rising sea levels like its neighbours. However,

Aotearoa is finally starting to realise that climate change is a real threat to the world's way of life and is making moves to becoming a net-zero country.

Our first steps have included the Zero Carbon Bill and new roles in Parliament like the Minister of Climate Change. But if we do not keep improving ourselves and holding ourselves accountable, our coastal communities will face rising sea levels, agriculture (one of our largest sectors) will suffer from a change in climate, our native way of life will suffer, and Aotearoa as we know it will change.

I wish our education system included more environmental and citizenship education. It is so incredibly important that the country is at least aware of what is happening, and then also aware of what it can do about it. Not nearly enough people enrol themselves to vote, let alone submit a ballot paper. I believe that if more people knew what was up, they'd be more inclined to act, and act strong. I also believe that people should know how their nation works, and that their vote means something.

Kia kaha. Stay strong. I know it feels terrible and overwhelming, like there is no one else out there who cares, but know that I am out there. The best time to act was when we first found out about climate change. The second best time is now.

ALEXANDER WHITEBROOK

Aged 25
AUSTRALIA

I am an activist for sustainable water management. Growing up in Western Australia, water scarcity was something I was aware of from a young age. I didn't realise it when I was younger, but little things such as not running the tap when brushing your teeth, or using a timer to keep your shower under four minutes long, were not normal things for many people in the world's richest and most water-abundant countries. This had a fundamental impact on me.

The preciousness with which we treated water in my home is something I want to share with the world. Years later, while living in Shanghai, I had the opportunity to be an intern at a small non-governmental organisation called Thirst, where I was inspired by my colleagues to put water conservation efforts at the centre of my career.

Water is an essential part of our everyday lives. No other resource is as complex or multifaceted as water, and so the United Nations itself has asserted it to be the 'primary medium through which we will feel the effects of climate change'. For these reasons, I put water at the forefront of my activism and advocacy against the climate crisis from my very first steps within this movement. Water pollution, droughts, floods, unpredictable rainfall, ice melt, change in river flows and mismanagement in industry and agriculture will increasingly threaten food security and political stability in the years to come.

* * *

Australia is going to be massively affected by the climate crisis. The 2019 summer was the hottest ever on record, and in December 2019 the country experienced its hottest national average daily temperature ever at 40.89 °C. While many people are taking action around the country, the political will to accelerate climate action is lacking.

If I could change one thing about my country tomorrow, I would put The Greens party in government. Since their founding in 1992 they have evolved from a one-issue micro-party to the third largest political party in Australia with rounded policies. The climate crisis takes precedence over all other issues in their political agenda, and that is exactly what Australia needs, to turn around the path of environmental destruction that it is currently on. I may not agree with all the political views of The Greens, but Australia must prioritise climate action the way they do.

The young people of the world are those who will be most affected by the increasingly severe consequences of the climate crisis. If anyone holds the highest authority over what we as a species must do to avoid disaster, it's us. The fact that we are still largely relegated to a token role in climate discussions is something we must continue to fight against.

Australia is the world's driest inhabited continent. About 35 per cent of the continent receives so little rain, it is effectively a desert.

KOMAL NARAYAN

Aged 27

FIJI

Pacific Islanders see the drastic impacts of climate change in their lives and the lives of members of their communities every day.

The biggest concern right now is rising sea levels: not just my country but all Pacific Islands countries are at risk of ending up underwater. In Fiji, three coastal communities have already been relocated so far due to this, and forty others have been identified as high-risk areas which will have to be relocated.

I realised that we young people are the key to this fight in climate change. I am part of a youth-led movement here in Fiji called the Alliance for Future Generations. We ensure that young people are represented equally at national, regional and international levels of climate advocacy. We also translate policies into common language for communities and conduct clean-up campaigns and mangrove planting activities with young people and adults.

Young people from all over the world are doing everything in their capacity to make a difference and fight the climate emergency. But the biggest challenge is being able to influence leaders and decision-makers. I am afraid that young people will not even have a future to lead.

Fiji is already vocal in its demands for urgency in the fight against climate change. The leaders are already actively asking world leaders to get serious about it. But until developed nations step up and cut emissions, Fiji alone won't be able to save itself.

We need to walk the talk and fight the fight.

KAILASH COOK

Aged 17
AUSTRALIA

For the first ten years of my life, I lived on a small island in the gulf of Thailand called Koh Tao, which is famous for marine tourism because of the local coral reefs. By snorkelling and swimming on these reefs, I was able to experience their complexity and beauty and marvel at the intricate weaving of nature. Simple things, like the symbiotic relationship between the clownfish and the anemone, or the cooperation of cleaner wrasse and groupers, were what really sparked my passion – such an array of different creatures all living together in harmony in the amazing ecosystem.

But I was also able to see the effects of climate change and how coral reefs were being threatened by human behaviour. In 2010, a bleaching event broke all previous temperature records, killing many of the reefs that I had grown up with. I remember a field of branching coral that stretched as far as the eye could see. I would always find beautiful fish amongst the branches of the coral and the dazzling array of life it supported. Then, in a period of about two months it was all gone. Just a field of rubble and sand.

The following year, in 2011, Koh Tao experienced intense rainstorms and flooding. Sediment run-off stifled the reef's recovery. These climate events, which have been taking place throughout my life, have been out of season and far more devastating than ever before. The examples clearly demonstrate the defining issue of our time. The realisation of the existential threat posed by climate change had a profound effect on my life and I felt, in some way, that I was responsible for these climate events that were destroying the ecosystems I loved so much.

My campaign is deeply personal: to channel my contribution towards climate change and simultaneously inspire and influence as many people as I can to do the same. My methods involve leading by example. I present my thoughts and ideas in a way that is designed to inspire people to create change in their own lives. Through my experiences, I aim to encourage people who are influenced by stories of hope and kindness rather than anger or fear.

Treating people with respect is always the most effective way to communicate.

The most challenging aspect of being a youth activist is the restrictions placed on young people. I've struggled with being able to attend events, visit facilities or participate in activities because of my age. In Thailand, regulations were less inhibiting and this freedom enabled me to operate on the front line, actively working to assist the recovery of our local reefs as an example to others. Since arriving in Australia in 2015, opportunities to participate in such activities have been far more limited.

I'm currently living in Townsville, Queensland, and in 2016 and 2017 the Great Barrier Reef experienced two consecutive bleaching events that wiped out many of the reefs that made up this phenomenal structure. Most greatly impacted were the highly restricted northern reefs, the ones least likely to be affected by humans. Or so we thought.

In 2019 we experienced an intense tropical storm that flooded the Townsville region. The excess water resulted in thousands of homes being destroyed and hundreds of millions of dollars' worth of damage. Seeing as this flood succeeded a seven-year drought in the region, it made the extreme weather event all the more unusual. I am only going to have to live through more of these extreme climate events, if climate change continues on its current trajectory.

If I could change one thing about Australia, it would be to convince everyone that climate change is real and must be acted upon. It pains me to see people suffering the effects of floods and fires in the same year who still think that climate change is propaganda. We still gain most of our international income from the export of fossil fuels, and our politicians refuse to act on the ever-pressing matter of climate change. Let's focus on turning Australia into a global leader in green energy and sustainability. Australia needs to lead by example so that the rest of the world may follow.

We all have to do two things. Firstly, we have to play our part. Ensure that you are taking responsibility for your own actions to limit carbon emissions where possible – whether that's limiting your electricity usage, using energy-efficient transport such as bicycles or public transport, or excluding carbon-heavy food from your diet. Secondly, we have to educate people on the issue: show them how they can contribute to lessening our impact on the climate. It's only when we work together that we can create a movement that will spark the changes needed for a sustainable future.

If global warming is limited to 1.5 °C, the Great Barrier Reef could survive. But at 2 °C, most scientists believe that 99 per cent of all the world's corals will be dead.

WHAT WOULD YOU SAY TO THE

We need to do our part and pull our weight. Please guide us into the next age safely, and keep everyone in mind when you make your next moves.

Lourdes Faith Auhura Parehuia,
18, New Zealand

People are concerned about their jobs and economic security, but none of this will be possible in the long term if we do not have climate action now.

Freya May Mimosa Brown, 17, Australia

LEADERS OF YOUR COUNTRY?

Developed nations are exploiting the Pacific. Our leaders need to stand up and say enough is enough, and protect our oceans and resources.

Komal Narayan, 27, Fiji

MADELEINE KEITILANI ELCESTE LAVEMAI

Aged 22

TONGA

In February 2018, Cyclone Gita hit Tonga and it was the country's most destructive natural disaster. The Category 4 to Category 5 cyclone destroyed buildings and homes, including that of my grandmother.

Witnessing natural disasters such as Cyclone Gita, and the severity of its impact on the country, tells me that there could be more cyclones with a worse impact on Tongan families and communities. Also, driving past one of the rural villages in Tonga, I noticed how the sea was getting closer to the roads. The sea will keep rising, putting people's homes and livelihoods in jeopardy.

The messages of Pacific Islanders have been overshadowed by those of activists in richer and more powerful countries. We also lack the resources needed to keep the movement alive and get our message across. When I first began campaigning, many activists relied on their own finances and financial support from their families.

The government should provide a platform for young and older activists, to amplify their voices. The grass-roots approach, for me, should be prioritised by leaders – this could include facilitating workshops for interested people, banning the use of plastics and encouraging people to live a more sustainable life. In 2018 the Vanuatu government made a bold move and banned single-use plastics and the whole country has got behind it – Tonga should do the same.

The government should not be advocating for more climate action while leaving the people out of their plans. Every person has a role to play in this climate crisis, and leaders should make that possible.

FREYA MAY MIMOSA BROWN

Aged 17
AUSTRALIA

I am incredibly lucky in that my parents are very supportive of my work. They have fostered in me a love of learning, a care for the environment, and the strength to stand up for what I believe in; they have been really influential in a positive way for my campaigning. This past year it has been challenging to balance activism with school, with sport, with family and friends, and my parents have helped me maintain that balance.

I have always cared deeply for the environment. Growing up, I began to learn more and more about the gravity of the crisis we are facing. Over the years I have been to many protests and signed many petitions, but I wanted to do more. At the beginning of 2019, I heard about a climate summer leadership program, which was a two-day introduction to youth environmental activism camp. After that, I joined the Melbourne School Strike 4 Climate team.

As part of the School Strike 4 Climate Australia movement, we are campaigning for climate action from our government. We have three demands:

1. No new coal, oil and gas projects, including the Adani mine.
2. 100 per cent renewable energy generation and exports by 2030.
3. Fund a just transition and job creation programme for all fossil fuel workers and communities.

Personally, I am particularly passionate about environmental economic sustainability and climate justice. It is particularly relevant in Australia, with its First Nations people. If I could change one thing about Australia, it would be our relationship with First Nations people. These are the first people to feel the worst impacts of the crisis. They also have so much experience and knowledge of the land and how to maintain it, which we could learn from. But there is little to no recognition, respect or support for indigenous communities. In conversations with indigenous people I have been shocked and saddened at how profoundly they are being affected.

* * *

Australia is feeling the impact of climate change largely through droughts and bush fires, which are having devastating effects on rural and farming communities. Currently, we are also very dependent on the fossil fuel industry, so anyone working in this sector is also feeling the effects as their job is at stake, which is why a just transition is so important.

Perhaps the most poignant singular moment that left a mark on me was going on a family camping trip and having a conversation with a farmer. He said that if there was no rain within the next week, they would lose their livelihood. Then on the way home we saw the Murray–Darling river basin almost completely dry. It can be easy to disconnect the issue from the individual people suffering, especially living in the city, and these conversations have been really important to ground me in the humanity of activism.

I think the most important thing is to remain hopeful. What we are facing is incredibly difficult, but it is not impossible. We must remain hopeful, as that is what is necessary to drive us to make change.

CARLON ZACKHRAS

Aged 19

MARSHALL ISLANDS

Student Carlon Zackhras gave this speech at the COP 25 meeting in Madrid, Spain, on 9 December 2019.

Iakwe and greetings from the Marshall Islands.

Before I came to Madrid, exactly two weeks ago, I experienced sixteen-foot swells that forced 200 people from their homes. Not only do we have inundations, but we also have epidemics of dengue fever, the flu, and I know our neighbours in Samoa are currently battling the measles that has taken seventy lives, thirty of which were children under the age of four. These are all illnesses linked to and made worse by climate change.

We've been told that if we want to stay living in our islands, we'll have to adapt and elevate our homes, with migration as the only plan B. We're having to deal with issues that we in the first place did not create. May I remind you that the Marshall Islands' contributions to climate change are only 0.00001 per cent of the world's emissions. My home is only two metres above sea level. With the threats of climate change we'd lose two metres of our culture, our *manit* our *iakwe*, our *roro*, our *biit* – two metres of our language, two metres of our legends.

This is why we created a Youth Leaders Coalition that enables young people to come up with innovative ways to adapt to climate change and lobby their ideas to our leaders. A team of students came up with new sea wall designs that our Minister of Environment David Paul has shown interest in. This is why our youth should be more involved. When you're in trouble, you find new solutions to the problem, and our youth has done just this.

This is why I'm here. To represent them and their ideas. To be a part of the generation that's going to end the fight. To voice out the reality of the future that the Marshall Islands is facing. To tell you that we don't want to lose our two metres.

ABOUT THE BOOK

The world adores heroes, but there are always more heroes than the few receiving attention. This book began as an essay making that point, published in September 2019 in *Quartz*, where I was a senior reporter covering climate change. Georgina Laycock, John Murray's publisher, asked me to find a way to feature those other heroes.

The result is this book, which started with the goal of capturing the stories of as many young climate activists as possible from as many different countries, backgrounds and experiences as we could find. Working with Abigail Scruby, an assistant editor at John Murray, I put together a list of nearly two hundred activists from more than a hundred countries and wrote to them.

The Instagram generation did not disappoint. I received many more replies than expected and they came sooner than I thought they would. These heroes are all fighting for the same cause, but I hope the book shows that they have many different reasons for protesting against the injustices that climate change is unleashing on the world. The unity in differences is what gives the movement its strength and momentum.

NOTES

ASIA
Fact sheet, page 8

one in seven Bangladeshis: 'Climate displacement in Bangladesh', Environmental Justice Foundation (2020), https://ejfoundation.org/reports/climate-displacement -in-bangladesh

more than 640 million people: International Monetary Fund, 'Boiling Point', *Finance & Development*, vol. 55, no. 3 (September 2018), https://www.imf.org/ external/pubs/ft/fandd/2018/09/southeast-asia-climate -change-and-greenhouse-gas-emissions-prakash.htm

Between 1998 and 2012: Ellen Gray, 'NASA finds drought in eastern Mediterranean worst of past 900 years', NASA (1 March 2016), https://www.nasa.gov/

feature/goddard/2016/nasa-finds-drought-in-eastern
-mediterranean-worst-of-past-900-years

More than a billion people: 'Reducing the impacts of
climate change', WWF – World Wide Fund for Nature
(2020), https://wwf.panda.org/knowledge_hub/where_
we_work/eastern_himalaya/solutions2/climate_change_
solutions/

Tatyana Sin, page 19
the Aral Sea: Dene-Hern Chen, 'The country that
brought a sea back to life', BBC Future (23 July 2018),
https://www.bbc.com/future/article/20180719-how-
kazakhstan-brought-the-aral-sea-back-to-life

Iman Dorri, page 24
The US recently renewed sanctions: 'Six charts that
show how hard US sanctions have hit Iran', BBC News
(9 December 2019), https://www.bbc.co.uk/news/
world-middle-east-48119109

Howey Ou, page 27
China is the world's largest emitter: 'China',
Climate Action Tracker (2 December 2019) https://
climateactiontracker.org/countries/china/

Liyana Yamin, page 35
100,000 premature deaths: Leah Burrows, 'Smoke from

2015 Indonesian fires may have caused 100,000 premature deaths', Harvard John A. Paulson School of Engineering and Applied Sciences website (19 September 2016) https://www.seas.harvard.edu/news/2016/09/smoke-2015-indonesian-fires-may-have-caused-100000-premature-deaths

NORTH AMERICA
Fact sheet, page 46

Parts of Canada's Arctic Ocean: 'Canada warming twice as fast as the rest of the world, report says', BBC News (3 April 2019), https://www.bbc.co.uk/news/world-us-canada-47754189

more than 140 million Americans: 'Health effects of ozone and particle pollution', American Lung Association (23 March 2020), http://www.stateoftheair.org/health-risks/

Almost all of the Caribbean's main cities: 'Can you imagine a Caribbean minus its beaches? It's not science fiction, it's climate change', The World Bank (5 September 2014), https://www.worldbank.org/en/news/feature/2014/09/05/can-you-imagine-a-caribbean-minus-its-beaches-climate-change-sids

extreme heat days: 'The effects of climate change', NASA (16 March 2020), https://climate.nasa.gov/effects/

Karel Miranda Mendoza, page 53
enabling 3 per cent of the world's maritime trade: 'Climate change threatens the Panama Canal', *The Economist* (21 September 2019), https://www.econ-omist.com/the-americas/2019/09/21/climate-change-threatens-the-panama-canal

more than 10 per cent: 'Panama economic outlook', Focus Economics (17 March 2020), https://www.focus-economics.com/countries/panama

Anya Sastry, page 61
The US has more than 70,000 miles of pipelines: Emily Moon, 'After the latest leak in South Dakota, how safe are America's pipelines?', *Pacific Standard* (24 November 2017), https://psmag.com/environment/how-safe-are-americas-pipelines

Cricket Guest, page 68
Métis are one of three: 'National Indigenous Peoples Day … by the numbers', Statistics Canada (20 June 2018), https://www.statcan.gc.ca/eng/dai/smr08/2018/smr08_225_2018

SOUTH AMERICA
Fact sheet, page 104

The Amazon absorbs a quarter: Josh Gabbatiss, 'Amazon carbon sink could be "much less" due to lack of soil nutrients', Carbon Brief: Clear on Climate (5 August 2019), https://www.carbonbrief.org/amazon-carbon-sink-could-be-much-less-due-to-lack-of-soil-nutrients

More than 99 per cent: Pool Aguilar León, 'Climate change and health in South America', Global Climate & Health Alliance (2018), http://climateandhealthalliance.org/resources/impacts/climate-change-and-health-in-south-america/

98 per cent of Andean glaciers: Jonathan Moens, 'Andes meltdown: new insights into rapidly retreating glaciers', Yale Environment 360 (30 January 2020), https://e360.yale.edu/features/andes-meltdown-new-insights-into-rapidly-retreating-glaciers

90 per cent of South America: Jorge Familiar, 'Climate change impacts in Latin America and the Caribbean: confronting the new climate normal', The World Bank (transcript dated 2 December 2014), https://www.worldbank.org/en/news/speech/2014/12/02/climate-change-impacts-in-latin-america-and-the-caribbean-confronting-the-new-climate-normal

Daniela Torres Perez, page 111
nearly 30 per cent: 'Peruvian glaciers have shrunk by 30 percent since 2000', Yale Environment 360

(7 October 2019), https://e360.yale.edu/digest/peruvian
-glaciers-have-shrunk-by-30-percent-since-2000

Catarina Lorenzo, page 113
the Pantanal wetlands: 'Brazil wildfires: Blaze
advances across Pantanal wetlands', BBC News (1
November 2019), https://www.bbc.co.uk/news/world
-latin-america-50257684

EUROPE
Fact sheet, page 128

100 times more likely: Daisy Dunne, 'Climate change
made Europe's 2019 record heatwave up to "100 times
more likely"', Carbon Brief: Clear on Climate (2
August 2019), https://www.carbonbrief.org/climate-
change-made-europes-2019-record-heatwave-up-to-
hundred-times-more-likely

Droughts and lower crop yields: Anouch Missirian
and Wolfram Schlenker, 'Asylum applications respond to
temperature fluctuations', *Science*, vol. 358, no. 6370
(December 2017), pp. 1610–14, https://science.
sciencemag.org/content/358/6370/1610

The number of wildfires: Alice Tidey, 'There have
been three times more wildfires in the EU so far this
year', Euronews (15 August 2019), https://www.

euronews.com/2019/08/15/there-have-been-three-times-more-wildfires-in-the-eu-so-far-this-year

This area could increase: Marco Turco et al., 'Exacerbated fires in Mediterranean Europe due to anthropogenic warming projected with non-stationary climate-fire models', *Nature Communications*, vol. 9, no. 3821 (2018), https://doi.org/10.1038/s41467-018-06358-z

heavier rainfall: 'Infographic: How climate change is affecting Europe', European Parliament News (20 September 2018), https://www.europarl.europa.eu/news/en/headlines/society/20180905STO11945/info-graphic-how-climate-change-is-affecting-europe

Raina Ivanova, page 143
Al Gore's documentary: Steven Quiring, 'Science and Hollywood: a discussion of the scientific accuracy of *An Inconvenient Truth*', *GeoJournal*, vol. 70, no. 1–3 (September 2007), https://doi.org/10.1007/s10708-008-9128-x

Federica Gasbarro, page 145
Climate impacts like drought: Anouch Missirian and Wolfram Schlenker, 'Asylum applications respond to temperature fluctuations', *Science*, vol. 358, no. 6370 (December 2017), pp. 1610–14, https://science.sciencemag.org/content/358/6370/1610

Adrián Tóth, page 158
Plastic contributes to greenhouse gas emissions:
Sandra Laville, 'Single-use plastics a serious climate
change hazard, study warns', *The Guardian* (15 May
2019), https://www.theguardian.com/environment/
2019/may/15/single-use-plastics-a-serious-climate-
change-hazard-study-warns

AFRICA
Fact sheet, page 162

Lake Chad: Will Ross, 'Lake Chad: Can the vanishing
lake be saved?', BBC News (31 March 2018), https://
www.bbc.co.uk/news/world-africa-43500314

droughts are projected to increase: 'Africa is
particularly vulnerable to the expected impacts of global
warming', United Nations Fact Sheet on Climate
Change (2006), https://unfccc.int/files/press/back-
grounders/application/pdf/factsheet_africa.pdf

Erosion rates: 'West Africa's coast losing over $3.8
billion a year to erosion, flooding and pollution', The
World Bank (14 March 2019), https://www.worldbank.
org/en/region/afr/publication/west-africas-coast-losing
-over-38-billion-a-year-to-erosion-flooding-and
-pollution

The continent recorded: Kiran Pandey, '195% more Africans affected due to extreme weather events in 2019', Down to Earth (29 December 2019), https://www.downtoearth.org.in/news/climate-change/195-more-africans-affected-due-to-extreme-weather-events-in-2019-68573

Toiwiya Hassane, page 186
The entire population: 'Addressing climate change in Comoros and Sao Tome and Principe', United Nations Economic Commission for Africa (4 September 2014), https://www.uneca.org/stories/addressing-climate-change-comoros-and-sao-tome-and-principe

Koku Klutse, page 189
Indoor air pollution: 'Household air pollution and health', World Health Organisation (8 May 2018), https://www.who.int/news-room/fact-sheets/detail/household-air-pollution-and-health

Tsiry Nantenaina Randrianavelo, page 192
climate change threatens: '25% of Madagascar's species threatened by climate change', WWF – World Wide Fund for Nature (15 March 2018), https://wwf.panda.org/?325358/25-des-especes-de-Madagascar-menacees-dextinction-par-le-changement-climatique

Delphin Kaze, page 199
Nine out of ten Burundians: Ministry of Foreign
Affairs of the Netherlands, 'Climate change profile:
Burundi', official report (April 2018), https://reliefweb.
int/sites/reliefweb.int/files/resources/Burundi_1.pdf

ANTARCTICA
Fact sheet, page 210

warming three times as fast: Louisa Casson, 'What
does climate change mean for the Antarctic?',
Greenpeace (blog dated 8 October 2018), https://www.
greenpeace.org.uk/news/what-climate-change-means
-for-the-antarctic/

Antarctic glaciers are losing ice faster: Diana
Madson, 'Are Antarctica's glaciers losing or gaining ice?',
Yale Climate Connections (8 August 2018), https://
www.yaleclimateconnections.org/2018/08/are-antarcti-
cas-glaciers-losing-or-gaining-ice/

emperor penguins: 'Whales, penguins and krill feel
the heat in Antarctica', WWF – World Wide Fund for
Nature (21 October 2019), https://www.wwf.org.au/
news/news/2019/whales-penguins-and-krill-feeling-the
-heat-in-antarctica#gs.vmxxvr

Krill: Institute for Marine and Antarctic Studies, 'More than just whale food: krill's influence on carbon dioxide and global climate', Science X (18 October 2019), https://phys.org/news/2019-10-whale-food-krill-carbon-dioxide.html

Over-fishing: 'Overfishing', Discovering Antarctica, https://discoveringantarctica.org.uk/challenges/sustain-ability/overfishing/

OCEANIA
Fact sheet, page 222

In 2019: 'Media reaction: Australia's bushfires and climate change', Carbon brief: Clear on Climate (7 January 2020), https://www.carbonbrief.org/media-reaction-australias-bushfires-and-climate-change

western Pacific sea levels: Alice Klein, 'Eight low-lying Pacific islands swallowed whole by rising seas', *New Scientist* daily newsletter (7 September 2017), https://www.newscientist.com/article/2146594-eight-low-lying-pacific-islands-swallowed-whole-by-rising-seas/

Coastal erosion: 'How Fiji is affected by climate change', COP23 Presidency website, https://cop23.com.fj/fiji-and-the-pacific/how-fiji-is-affected-by-climate-change/

two-thirds of New Zealand's population:
'Flooding', Climate Change Implications for New
Zealand key risks, Royal Society of New Zealand / Te
Apārangi, https://www.royalsociety.org.nz/what-we-do
/our-expert-advice/all-expert-advice-papers/climate
-change-implications-for-new-zealand/key-risks/flood-
ing/

Alexander Whitebrook, page 228
the world's driest inhabited continent: 'Deserts',
Geoscience Australia (2020), https://www.ga.gov.au/
scientific-topics/national-location-information/land-
forms/deserts

Kailash Cook, page 233
If global warming is limited: National Oceanic and
Atmospheric Administration (NOAA), US Department
of Commerce, 'Four-month coral bleaching outlook',
Coral Reef Watch, https://coralreefwatch.noaa.gov/
satellite/bleachingoutlook_cfs/outlook_cfs.php

USEFUL RESOURCES

In one way, climate change is a simple phenomenon. The burning of fossil fuels is adding greenhouse gases to our atmosphere, which is causing the global average temperature to rise. But what happens when warming occurs isn't always easy to predict.

Similarly, tackling climate change is straightforward: lower the amount of greenhouse gases pumped into the atmosphere. But because so much of our lives remain dependent on fossil fuels, transitioning away is complicated.

Below you'll find a selected list of resources that provide information about climate science, charities that fight climate change, organisations that sell carbon offsets, and activist groups pushing for change.

Science

NASA's Global Climate Change at www.climate.nasa.gov

Climate Central at www.climatecentral.org
Intergovernmental Panel on Climate Change www.ipcc.ch

Charities
Coalition for Rainforest Nations www.rainforestcoalition.org
Ocean Conservancy www.oceanconservancy.org
Project Drawdown www.drawdown.org

Carbon offset
United Nation's carbon offset platform at offset.climateneutralnow.org
Cool Effect at www.cooleffect.org
Gold Standard at www.goldstandard.org

Activism
Fridays For Future at www.fridaysforfuture.org
350.org at www.350.org
Extinction Rebellion at www.rebellion.earth

ACKNOWLEDGEMENTS

This book wouldn't be a reality without a lot of people's hard work. Georgina Laycock spotted the idea, and then encouraged me to pursue it. Abigail Scruby was the book's editor, and she helped me through all of the little difficulties that came in the way of putting together this complex jigsaw puzzle. Jonathan Conway, my agent, helped me to wade through the many knotty problems that came up in the publishing process. And Deeksha, my wife, gave me the energy to work through long stretches without breaks.

Most of all, however, I'd like to thank all the young activists and their parents for allowing us to share these stories more widely. The long list below names every person who sent in their story, even though we couldn't feature every entry.

Aditya Mukarji

Adrián Tóth

Agim Mazreku

Akari Tomita

Alberto Barrantes Ceciliano

Albrecht Arthur N. Arevalo

Alexander Whitebrook

Alexandros Nicolaou

Anna Taylor

Anya Sastry

Arpan Patel

Ashley Torres

Ayakha Melithafa

Boubacar Mahamadou
  Maiga

Brandon Nguyen

Brishti Chanda

Carlon Zackhras

Catarina Lorenzo

Cecilia La Rose

Chiara Sacchi

Côme Girschig

Cricket Guest

Daniela Torres Perez

Delphin Kaze

Dilangez Azizmamadova

Elijah McKenzie-Jackson

Elizabeth Wanjiru Wathuti

Emma-Jane Burian

Emmanuel Lobijo Josto Eka

Eva Astrid Jones

Eyal Weintraub

Federica Gasbarro

Freya May Mimosa Brown

Gilberto Cyril Morishaw

Holly Gillibrand

Howey Ou

Htet Myet Min Tun

Iman Dorri

Iryna Ponedelnik

Jamie Margolin

Jeremy Raguain

Jiaxin Zhao

João Henrique Alves
  Cerqueira

John Paul Jose

Juan José Martín-Bravo

Kailash Cook

Kaluki Paul Mutuku

Karel Lisbeth Miranda
  Mendoza

Khadija Usher

ACKNOWLEDGEMENTS

Koku Klutse
Komal Narayan
Laura Lock
Lesein Mathenge Mutunkei
Lia Harel
Lilith Electra Platt
Liyana Yamin
Lourdes Faith Auhura
  Parehuia
Lucie Smolkova
Mackenzie Feldman
Madeleine Keitilani Elceste
  Lavemai
Maja Starosta
Mareeka Dookie
Maryam Kharusi
Nadine Clopton
Nasreen Sayed
Nche Tala Aghanwi
Ndéye Marie Aida
  Ndieguene
Nijat Eldarov
Octavia Shay
  Muñoz-Barton

Payton Mitchell
Pierre Garcia
Pramisha Thapaliya
Raina Ivanova
Raslen Jbeli
Ricardo Andres Pineda
  Guzman
Rimante Balsiunaite
Ruby Sampson
Santiago Aldana
Sebenele Rodney Carval
Shannon Lisa
Stamatis Psaroudakis
Tafadzwa Chando
Tatyana Sin
Theresa Rose Sebastian
Toiwiya Hassane
Tsiry Nantenaina
  Randrianavelo
Vania Santoso
Vishnu P R
Vivianne Roc
Zoe Buckley Lennox

ABOUT THE CONTRIBUTORS

Aditya Mukarji (16) is a student from India who began campaigning against single-use plastic at the beginning of 2018. His campaign #RefuseIfYouCannotReuse has since removed over 26 million straws and several million other single-use plastics from the hospitality industry. In 2019 he was invited to attend the UN Youth Climate Action Summit.
Twitter: @AdityaMukarji

Htet Myet Min Tun (18) is from Yangon, Myanmar, and has worked at the Myanmar Institute of Strategic and International Studies. At his school he led an environmental conservation project and engaged the local community to increase public awareness of climate change. In 2019 he was invited to attend the UN Youth Climate Action Summit.

htetmyetmintun.com LinkedIn: linkedin.com/in/htet-myet-min-tun-728108185

Tatyana Sin (26) is from the Khorezm region of Uzbekistan, near the Aral Sea. She recently completed her master's degree. She promotes environmental awareness and has previously worked at the UNESCO Tashkent office and the Global Environment Facility Small Grants Programme.

Iman Dorri (28) is an environmental engineer from Iran. His work focuses on sustainable development. He has an MSc in environmental engineering from the Amirkabir University of Technology and works for the university's office of sustainability.
Instagram: @imandorri LinkedIn: linkedin.com/in/iman-dorri-714844135/

Howey Ou (17) is a vegan and climate activist calling for climate action in China. She is the founder of the #PlantforSurvival movement. She travelled around China alone and without financial support, protesting outside government buildings and visiting climate and environmental NGOs.
Twitter: @Howey_Ou

Theresa Rose Sebastian (16) is a student living in Ireland. She has organised several protests to call for national action on climate change. She has attended and spoken at conferences and strikes, and is passionate about raising the voices of people in the global south.

Nasreen Sayed (27) is a US-based environmental professional from Afghanistan. She is currently serving as a fellow with the Local Government Commission, building community resilience through the implementation of sustainable projects and strategies. She holds an MSc in environmental technology from Imperial College London. Facebook: /Nasreen.sayed.1460 Instagram: @nassayed

Liyana Yamin (27) is a PhD student at the National Taiwan Ocean University. She is actively involved with the Malaysian Youth Delegation, the only youth-led organisation in Malaysia, which focuses on climate-change policy.

Albrecht Arthur N. Arevalo (26) is a Filipino youth leader working with non-government organisations, faith-based groups and government. He campaigns for the UN Sustainable Development Goals and is a firm believer in community based and collaborative action. Facebook: /Albrecht.arevalo Instagram: @brexarevalo

Akari Tomita (16) is a US-based student campaigning for youth representation in climate action and zero waste. She is a member of her school's environmental club and has spoken as a panel member in a sustainability briefing at the UN.

Cecilia La Rose (16) is a student from Canada. Her activism has involved attending and organising protests, speaking with politicians in her community about the climate crisis and, most recently, she was a co-plaintiff in a lawsuit filed against the Canadian federal government over their inaction and contribution to the climate crisis.

Karel Lisbeth Miranda Mendoza (27) is a Panamanian biologist, leader and climate activist. She is the founding member of the organisation Youth Against Climate Change in Panama and currently vice-president of its board of directors.
Instagram: @karelssy Twitter: @karell_lissy

Emma-Jane Burian (18) is a 2020 Victoria Leadership Award winner, grade-12 student and climate activist who organises global and monthly climate strikes with Our Earth, Our Future on the beautiful homelands of the Lekwungen and W̱SÁNEĆ peoples in Victoria, BC. She is passionate about climate justice and engaging youth in politics to build resilient and strong communities for a better world.
Instagram: @emmajanevictoria Twitter: @EJburian

Anya Sastry (18) is a student from Illinois. Her activism focuses on preventing politicians from taking fossil fuel money, combating further installation of fossil fuel infrastructure and campaigning to pass elements of the Green New Deal through Congress. She has coordinated regional and national protests with US Youth Climate Strike. She was invited to the UN Youth Climate Summit.

Ricardo Andres Pineda Guzman (22) is a Honduran sustainable-development advocate and a champion of decarbonisation. He is collaborating with officials to pair climate action with inclusive green finance so that the effects of climate change can be mitigated while also providing growth and development opportunities in Honduras, deemed the most climate vulnerable country of the world.
Instagram: @ricardopineda1
LinkedIn: /ricardopinedaguzman

Cricket Guest (22) is a film-maker, actress and activist. She works with Climate Strike Canada and is the indigenous outreach coordinator for Fridays for Future Toronto. She seeks to centralise indigenous voices in the climate crisis conversation.

Lia Harel (19) has previously advocated for and helped write climate action policies at local, state and national level. She is currently an undergraduate student at Claremont McKenna College in California majoring in environment, economics and politics.

Facebook: /lia.harel.7 Instagram: @lia_harel

Shannon Lisa (22) is a 'toxic detective' from New Jersey and the program director of the non-profit organisation Edison Wetlands Association that investigates the effects of chemical contaminant dumping in communities in Indiana and beyond. In 2019 she was one of six winners of the Brower Environmental Youth Awards.

Khadija Usher (26) is a researcher at the department of socio-environmental energy science at the Graduate School of Energy Science, Kyoto University, Japan. Her research focuses on transitioning to sustainable energy in small economies. Originally from Belize, she has spearheaded national projects such as Belize's Energy Planning and Strategy 2035.

Brandon Nguyen (20) is a Toronto-raised environmental activist currently pursuing his undergraduate education at the Wharton School of Business at the University of Pennsylvania. He is particularly interested in climate finance, urban sustainability and renewable energy policy.

Facebook: /brandon.b.n

Instagram: @brandon.nguyen1999

Vivianne Roc (22) is a third-year pharmacy student based in Haiti. She is the founder of an organisation called Plurielles which works primarily with young women and girls and seeks to integrate climate action into efforts to address public health.

Octavia Shay Muñoz-Barton (16) is a student and member of Heirs To Our Oceans, a global youth leadership organisation committed to protecting oceans and waterways, fighting for environmental justice and creating a global movement of youth who are empowered and empathetic leaders.
h2oo.org

Payton Mitchell (21) is a student and climate activist. She helped to start Climate Strike Canada and was a part of Emma Lim's #NoFutureNoChildren pledge. Currently she works with CEVES (the Quebec student coalition) to recruit students outside Quebec to join the network.

Ashley Torres (23) is a student and a spokesperson for the Quebec student environmental movement. She protests against fossil fuel projects and campaigns for environmental justice for indigenous communities.

Eyal Weintraub (20) is a student from Argentina. He is one of the founders of Jóvenes por el Clima Argentina

(JOCA) which has become the biggest youth climate movement in Argentina.

Daniela Torres Perez (18) is one of the co-founders of the UK Student Climate Network (UKSCN). In 2019 her writing was featured in the book Letters to the Earth: Writing to a Planet in Crisis.

Catarina Lorenzo (13) is a student from Salvador, Brazil. She was one of sixteen children to file a complaint to the UN Committee on the Rights of the Child to protest at lack of government action on the climate crisis.

Juan José Martín-Bravo (24) is a Chilean environmentalist dedicated to sustainability and entrepreneurship. He co-founded and leads Cverde, an environmental NGO. He was the first young negotiator for Chile at COP25 and was appointed as the general coordinator for the 15th Global Conference of Youth (COY15), a global conference under the UNFCCC, all while studying aerospace engineering.
Instagram: @juanjo.martinb   Twitter: @JuanjoMartinb

João Henrique Alves Cerqueira (27) is a bachelor's student in environmental engineering from Brazil. He founded the Curitiba climate coalition and has a project where people travel by bicycle to hear the stories of people at the forefront of the climate crisis, especially in

traditional and indigenous communities.
Instagram: @joaohencer

Gilberto Cyril Morishaw (25) is a master's student born in Curaçao and currently based in the Netherlands. In 2019 he launched a think tank about food security in the Dutch Caribbean for the Dutch Ministry of Agriculture. He is an ambassador for the African Caribbean Pacific Young Professionals Network where he focuses on fighting inequality.

Holly Gillibrand (15) is a student from Scotland. She is a volunteer for OneKind, a rewilding advocate and a young ambassador for Scotland: The Big Picture.
Twitter: @HollyWildChild

Stamatis Psaroudakis (22) is a university student from Greece. His activism focuses on intersectional LGBTQIA+ equality, raising environmental awareness and combating xenophobia. Stamatis was invited to join the EU Young Citizens Dialogue on the Fighting Climate Change Panel and attended the UN Youth Climate Summit in 2019.

Lilith Electra Platt (11) is an international environmental champion. She has been named one of the 100 influencers tackling plastic pollution. Lilly is a youth ambassador for the Plastic Pollution Coalition,

YouthMundus and WODI. She uses her social media platforms to raise awareness about environmental issues such as plastic pollution and climate change.
Instagram: @lillys_plastic_pickup Twitter: @lillyspickup

Anna Taylor (19) is a UK climate justice activist who founded the UK Student Climate Network when she was seventeen years old. Through this she has helped coordinate the youth strikes in the UK and across Europe as part of the Fridays For Future movement, pressurising the government to declare a climate emergency and advocating for maintaining mental well-being in the context of the climate crisis.
Instagram: @anna.e.taylor Twitter: @AnnaUKSCN

Raina Ivanova (15) is a student from Germany. She was one of sixteen children who petitioned the UN Committee on the Rights of the Child to protest against international government inaction over the climate crisis.

Federica Gasbarro (25) is an Italian climate activist, writer and student of biological science at the University of Rome Tor Vergata. She was invited to represent Italy at the UN Youth Climate Summit in 2019.
Facebook: /Federica.gasbarro
Instagram: @federica_gasbarro

Laura Lock (18) is a British-Hungarian student currently studying for the IB diploma in Oxford. After attending several school climate strikes she became fascinated by the role of indigenous groups and young people in the global climate conversation. She has worked with the Centre for United Nations Constitutional Research and the ONE Campaign as a youth ambassador.
Instagram: @laura.lock

Agim Mazreku (23) is from Kosovo and is currently studying for a master's in climate science and policy. He campaigns for a just and radical transformation of energy systems to renewable and clean sources.
LinkedIn: linkedin.com/in/agimmazreku

Adrián Tóth (30) is the co-founder of Plastic Free Plux in Brussels, an organisation which seeks to eliminate single-use plastics in Brussels.
Instagram: @plasticfreeplux Twitter: @plasticfreeplux

Kaluki Paul Mutuku (27) is a Kenya-based climate advocate and environmental defender with a background in environmental conservation and natural resource management from the University of Nairobi, Kenya. He has previously worked with the African Youth Initiative on Climate Change (AYICC), 350.org and A Rocha Kenya. He is the regional coordinator for the African

UN group Youth4Nature, a global youth-led and youth-serving organisation mobilising nature based solutions to climate action.
Facebook: /PrincePaulh Twitter: @KalukiPaul

Nche Tala Aghanwi (25) is a science diplomat, a sustainable development activist and political pundit. He is the founder and executive director of the Africa Science Diplomacy and Policy Network
(ASDPN). He has a master's in international cooperation, humanitarian action and sustainable development from the International Relations Institute of Cameroon (IRIC).
Facebook: /ASDPN.INTERNATIONAL   LinkedIn: linkedin.com/company/africa-science-diplomacy-and-policy-network-asdpn

Sebenele Rodney Carval (30) currently works for the Ministry of Tourism and Environmental Affairs in Eswatini as a climate finance project coordinator. He holds a degree in energy engineering and has a back-ground in sustainable energy.
Facebook: /rodney.carval Instagram: @rodneycarval

Jeremy Raguain (26) is a project officer for the Seychelles Islands Foundation. He holds a BSocSc in environmental geographical science and international relations, as well as an honours degree in international

relations from the University of Cape Town. He is a Sustainable Ocean Alliance young ocean leader as well as a global shaper from the Victoria Hub and is also the East African regional lead for the Global Shapers Climate and Environment Steering Committee.
Instagram: @turtlecommuter Twitter: @mahesituated

Lesein Mathenge Mutunkei (16) is a student from Kenya. He founded Trees for Goals, an organisation which aims to use football to raise awareness of the climate crisis and combat deforestation.
Instagram: @trees4goals Twitter: @trees4goals

Toiwiya Hassane (21) is a biology student and a member of the Indian Ocean Climate Network from Comoros. Her work is dedicated to the field of green entrepreneurship and promoting sustainable agriculture.

Koku Klutse (28) is the manager of Jony Group, a biogas promotion company. He is an environmental activist based in Togo, specialising in energy transition.
Facebook: /jano.klutse Instagram: @casimirklutse

Tsiry Nantenaina Randrianavelo (28) is a climate activist from Madagascar and founder of the Move Up Madagascar NGO which is focused on youth and climate. He holds an MBA degree and is a specialist in civic lead-ership and project management who has represented

Malagasy Youth to the UN Youth Climate Summit of 2019.
Facebook: /tsirynantenaina.randrianavelo
Twitter: @tsr135

Ruby Sampson (19) is a South African climate activist and co-founder of the youth-led Afrocentric organisation African Climate Alliance. She first witnessed the dramatic effects of climate change when travelling around Africa in a truck (africaclockwise.wordpress.com), and since then has been fully dedicated to climate education and creating a network of climate activists across Africa.
africanclimatealliance.org
Instagram: @africanclimatealliance

Tafadzwa Chando (23) is a youth and climate activist based in Zimbabwe, and also the founder of Magna Youth Action. He has been engaged in activism since he was fourteen.
Twitter: @tafaddwilliams

Delphin Kaze (25) is a Burundian social entrepreneur, innovator and climate activist. He has a BSc in environmental sciences (climate and biodiversity) from the Université Polytechnique de Gitega. He is the founder and CEO of KAze Green Economy (KAGE) Ltd, a social

enterprise that provides clean cooking energy in Burundi.
Facebook: /delphin.kaze Twitter: @kaze_delphin

Elizabeth Wanjiru Wathuti (24) is a Kenyan climate
activist and founder of the Green Generation Initiative
which aims to address global environmental challenges
and to educate young people about the environment. In
2019 she was awarded the Africa Green Person of the Year
Award by the Eleven Eleven Twelve Foundation and
named as one of the 100 Most Influential Young Africans
by the Africa Youth Awards.
Twitter: @lizwathuti

Ndéye Marie Aida Ndieguene (24) is a civil engineer,
climate activist, writer and entrepreneur from Senegal.
She builds storage spaces from recycled materials for
farmers, and leads other initiatives like Nawari, a platform promoting local and organic products made in
Africa, and Environnementalistes, a group of environmental experts implementing actions to fight climate
change in Senegal.
Instagram: @a__senegalese_writer

Zoe Buckley Lennox (26) is a climate activist based in
Brisbane, Australia. She has worked with Greenpeace over
the years and has gone on expeditions to the Arctic and
Antarctica to protest against oil drilling and krill fishing.

Lourdes Faith Auhura Parehuia (18) is a student from Auckland, New Zealand. She campaigns for climate action and indigenous rights, and is currently the Green Party candidate for Manurewa.
Facebook: /lourdesfvano Twitter: @lourdes_vano

Alexander Whitebrook (25) is an activist for sustainable water management and a board member for the World Youth Parliament for Water. His work has involved sitting on the steering committee of the Global Water Partnership and advising UN Water as part of its expert group on regional coordination.

Komal Narayan (27) is a climate change activist representing the voices of Fiji and the Pacific. She is the assistant coordinator for the Alliance for Future Generations Fiji. She is currently pursuing her MA in development studies, focusing on climate induced relocation in Fiji, and has previously been a Green Ticket recipient for the UN Youth Climate Action Summit and participated in COP23, COY13, COP25 and COY15.
Facebook: /Komal.kumar.750331
Instagram: @karishma.komal92

Kailash Cook (17) is an Australian high-school student with a passion for coral reefs and ocean ecosystems. He has participated in reef-restoration activities to improve the health of reefs around the world and presented research at conferences including the Great Barrier Reef Restoration Symposium and Plan for the Planet, Mauritius.

Madeleine Keitilani Elceste Lavemai (22) is a student and co-founder of Pacific Islands Students Fighting Climate Change (PISFCC). Currently she is advocating for support from Pacific Island nations to commence proceedings for an advisory opinion on human rights and climate change by the International Court of Justice. Twitter: @pisfcc

Freya May Mimosa Brown (17) is a student and climate activist involved in leading the Melbourne School Strikes for Climate in 2019. She is currently completing her final year of the International Baccalaureate, and looking to pursue a career in climate science research.
Facebook: /freyammbrown   Instagram: @freyamimosa

Carlon Zackhras (19) is a student from the Atoll Nation of the Marshall Islands. In 2019 he spoke at COP25 in Madrid about the threat that climate change poses to his home.
Twitter: @thejajok

INDEX

INDEX

INDEX

palm oil 35
Panama 53–6
Panama Canal 53–4, 56
Pantanal wetlands 114, 115
Parehuia, Lourdes Faith Auhura 225–7, 238
Paris Agreement 3, 27, 28
Paul, David 246
Peru 104, 111–12
petrol nationalism 96
Philippines 38–40
#PlantForSurvival 28
plastic 4, 42, 138, 158–9, 160
    and marine life 146
    and Seychelles 177
    and single-use 11–12, 241
Plastic Free Plux 159
Platt, Lilith Electra 138–9
Plurielles 89
poison gases 81
politics 13, 56, 151–3
    and Afghanistan 33–4
    and Australia 229–30
    and Iran 26
    and Kenya 167–8
    and lawsuits 49–50
    and Myanmar 16, 17
    and Seychelles 176–7
    and South Africa 195
    and Tonga 241
    and Zimbabwe 197–8
pollution 46
    and Curaçao 123–4
    and India 13
    and Panama 54
    and USA 62–3, 78–82
poverty 13, 19, 124, 162, 173–4

protests 24
    see also school strikes
Psaroudakis, Stamatis 135–7

Raguain, Jeremy 176–9, 184
rainfall:
    and Afghanistan 33
    and Australia 230
    and Brazil 114
    and Kenya 165
    and Malaysia 35
    and Myanmar 16
    and Panama 53–4
    see also flooding
rainforests 104, 121
Randrianavelo, Tsiry Nantenaina 192–3
recycling 11, 42
refugees 3, 147
#RefuseIfYouCannotReuse 11–12
religion 38–9
respiratory disease 21
reusable cups 158–9
rice 13
Roc, Vivianne 4, 88–90

salt storms 21
Sampson, Ruby 185, 194–6
sanctions 26
Sanders, Bernie 116
Sastry, Anya 61–5, 76
Sayed, Nasreen 23, 33–4
school strikes 1–2, 6, 150–5
    and Australia 242–3
    and Ireland 30, 31–2
    and Scotland 131, 132
Scotland 131–4, 148

INDEX